GERMAN ENGINEERS

The Anatomy of a Profession

GERMAN ENGINEERS

The Anatomy of a Profession

STANLEY HUTTON

AND

PETER LAWRENCE

CLARENDON PRESS · OXFORD
1981

Oxford University Press, Walton Street, Oxford OX2 6DP

London Glasgow New York Toronto
Delhi Bombay Calcutta Madras Karachi
Kuala Lumpur Singapore Hong Kong Tokyo
Nairobi Dar es Salaam Cape Town
Melbourne Auckland
and associate companies in
Beirut Berlin Ibadan Mexico City

Published in the United States by
Oxford University Press, New York

British Library Cataloguing in Publication Data

Hutton, Stanley
German engineers.
1. Engineering — Study and teaching —
West Germany
I. Title II. Lawrence, Peter
620'.004'071143 TA174
ISBN 0-19-827245-6

Typeset by Oxford Verbatim Limited
and printed in Great Britain
at the University Press, Oxford
by Eric Buckley,
Printer to the University

To
OUR WIVES

Preface

This is a general book about German engineers. It aims to give some account of their recruitment and training, their deployment and prospects within the occupational structure, and to discuss their position in German society and contribution to the achievements of German industry.

To some extent these over-all objectives are served by a particular source of information. At the centre of our data are the results of a detailed survey of a sample of just over 1,000 mechanical engineers in West Germany, a survey we undertook on behalf of the British Department of Industry, 'Recruitment, Deployment, and Status of the Mechanical Engineer in the Federal German Republic', published in 1977. The survey was based on a 215-item questionnaire, many of these items being multipart questions. The questionnaire was actually administered in interviews lasting one and a-half to two hours. This interviewing work was conducted by the staff of a German survey institute. The questionnaire in turn was a slighly modified version of that developed by Joel Gerstl and Stanley Hutton in a survey of mechanical engineers in Britain in the 1960s, and the sample-size and structure was very similar in the two surveys. Thus at some points in the book we directly contrast findings from the British and German surveys, subject to a cautionary word about the lapse of time between them.

In analysing the results of this German survey we are able to present the responses to the various questionnaire items not only for the whole sample but for various sub-groups as well. These latter include graduates (Dipl. Ings.) and non-graduates (Ing. Grads.), engineers who are members of the major professional association (the VDI), those who are members of a trade union, those who work in industry, those who work in the public sector, those in the top and bottom salary quartiles for their age group, and those with above average and below average degree or engineering-school qualification grades. In many cases there are interesting differences between the various sub-groups, and in consequence many references to these sub-group analyses in the following chapters.

There have also been many other inputs besides this survey. In the work which preceded the writing of this book the authors benefited from a familiarity with the relevant literature and some recent German research. We obtained information from several official sources, visited a range of educational and research institutes in West Germany, and were helped by discussions with German engineers, officials, and academics. One of the authors has also visited many industrial firms in Germany to discuss the training and deployment of engineers, and has also interviewed German managers and spent periods as an observer in a number of German companies.

It is important to underline this plethora of sources and inputs. In some sections of the book, in for instance the discussions of occupational recruitment and success, the use of the survey data is crucial. Others, the discussion of the German education system for example, are based primarily on different sources of data and understanding. Furthermore, some of the more general themes which are treated in the book, including our assessment of the standing of engineering in Germany and the status of industry, are informed by the totality of our knowledge and experience.

It is in these senses, the variety of inputs and a readiness to discuss the engineer's place in German society, that this is a general book rather than the account of a particular piece of survey based research. To put it another way, this book embraces more than a portrait of the German engineer: it is also in part a characterization of German society.

Acknowledgements

Much of the material for this book derives from a study of German engineers which we made for the British Department of Industry. We would like to thank the Department of Industry for their willingness to let us publish the results and insights from this study. In this connection we would also particularly like to thank Michael Fores, former Senior Economic Adviser at the Department of Industry, for his support and encouragement. The study involved a survey of mechanical engineers in West Germany, in which a lengthy and detailed questionnaire was completed in personal interviews. We are grateful to the survey institute Infratest of Munich for carrying out this field work under our direction, and to the 1,006 German engineers who were kind enough to allow themselves to be interviewed.

Especially in the early stages of the study we were assisted by the British Embassy in Bonn with advice, introductions, and help in scheduling many series of visits in Germany. We also benefited from discussions with many academics and administrators in Germany including: Dr Reuhl of the Sekretariat der ständigen Konferenz der Kultusminister der Länder in der Bundesrepublik Deutschland; Dipl. Ing. Heinz Justi of Siemens AG; Dr Hermann Bayer and Professor Heinz Hartmann of the Westfälische Wilhelms-Universität; Dr Karl Boetticher of the Institut für Wirtschafts-und Sozialforschung at Giessen; Professor Rudolf Quack of the Universität Stuttgart; Professor Werner Fister and Professor Heinz Fasol of the Ruhr-Universität Bochum; Dr Walter Müller and Dr Rainer Ruge of the Universität Mannheim; Dr Annelie Hummer of the Deutsches Institut für Internationale Pädagogische Forschung, Frankfurt; Dr Gerd Hortleder then of the Technische Universität Berlin; Dr Bernard Wilpert and Dr Arndt Sorge of the International Institute of Management, West Berlin; and with the late P. V. Vassen of the Deutsches Institut zur Förderung des industriellen Führungsnachwuchses in Cologne.

We have also variously received help, advice, and information from a number of organizations in West Germany including: the Verein Deutscher Ingenieure, whose consultant sociologist Holger Hillmer we would particularly thank; from the Union der Leitenden Angestellten and several of its member organizations; from the Deutsche Angestellten Gewerkschaft; and from the Bundesanstalt für Arbeit; and by post from the Statistisches Bundesamt and the Deutsches Gewerkschaftsbund.

During the course of the study we had the opportunity to visit a number of educational institutes where engineers are trained and in this connection we would like to thank academics, teachers, and administrators at the Fachhochschule Konstanz; the Hochschule für Technik Bremen; the Technische Universität Braunschweig; the Universität-Gesamthochschule Essen; the Technische Hochschule München; the Gesamthochschule Kassel and the Gesamthochschule Siegen.

Towards the end of the study we also visited many manufacturing companies in West Germany and benefited greatly from discussions with personnel managers, training officers, and other executives about both the deployment of engineers and the general nature of industrial management in Germany. These companies are perhaps too numerous to name but our thanks to them is very real.

Finally we would like to thank Mrs Diana Hodgkinson for a fast and fluent typing of the Manuscript.

University of Southampton   S. P. HUTTON
February 1980   P. A. LAWRENCE

Contents

Chapter 1

Introduction

Just across the road from Central Station in Amsterdam there is a sign which reads:

> 5,397 miles to Wall Drug,
> Wall, South Dakota.

It is not recorded whether this sign was erected by a homesick GI or an idiosyncratic tourist, but the interesting thing is that although it states a geographic fact it expresses a social one. What the sign really means is that Holland and the USA are different: the actual distance is not significant (after all, it is a long way from Texas to South Dakota too) but the sense of not being at home is.

The idea that there may be real differences between countries is important, and not obvious. Much writing about the modern world and much common observation, tends to support the other view that everywhere is much like everywhere else. Among the rich industrial countries, in particular, one is accustomed to note similarities. There are the same motorways and petrol stations, goods and advertisements, even fears and aspirations; they all have engineers, managers, manufacturing industry, apprentices, technicians, professional associations, schools and school subjects, technical colleges and universities, and 'of course' they are all the same. If, indeed, one is convinced about international homogeneity, 'comparison' becomes something of a self-fulfilling operation. The technique is to take one's own country as *point de départ* and simply shuffle around with the data on the compared country till a matching set of forms and institutions is found. The trouble is that this 'matching set' may not really be the same; the relativities in the other country may be different, and people may take a different view of entities which have the same label.

This book, however, is written in the conviction that there may well be substantial differences between countries, even between countries in the richer, industrialized, free and democratic group. This represents not only a systematic interest, but is a *raison d'être* for these pages. It is precisely because German engineers occupy a qualitatively different status in their own society that they are worth writing about. This is not the only reason, but from the standpoint of this side of the North Sea, it is a powerful one. German engineers contribute a lot to their society: they are clearly able to, and they are also encouraged to.

Putting this basic idea another way, this book is about German engineers, but both the authors are British and so is their viewpoint. So we are interested not only in the way things are in Germany, but also in the differences, the reasons behind them, in emerging configurations, and indeed in anything practical one can learn from all this.

Apart from this comparative interest, and its implications for wider characterization, there are many things which render engineers worthy of a sympathetic monograph — in any national context. As an earlier commentator has pointed out, it is surprising how little is really known about engineers.[1] As a white-collar occupational group they are superseded in numbers only by school teachers, yet it is difficult to think of half a dozen books in English about them.

This relative neglect in the medium-length literature is made more poignant by the topicality which engineers have enjoyed in Britain in the late 1970s. This interest is, of course, partly extrinsic. Britain has done poorly economically, and concerned discussion has explored a variety of determinants and contributory factors including the role of the engineer.[2]

The interest in engineers and engineering shown in Britain in the last few years, however, goes beyond the economic performance debate and has achieved some momentum of its own. Conferences and seminars on the training and utilization of engineers have mushroomed, and research on engineering themes, in this sense, has become common. In 1977 the British Association commissioned a substantial study of British engineers as a discussion spring board for the Annual Conference of that year,[3] and the 1978 Annual Conference included for the first time a section on the sociology of engineering. At the same time, there has been a growing interest in engineers in other countries, with newspapers and journals offering occasional articles and sometimes systematic series on engineers abroad.[4] The Finniston Committee was established to inquire into the training and employment of British engineers just after the empirical work on which this book is based was completed and the Finniston Committee's Report appeared while the present book was being written.

If topicality is a further justification for a monograph on engineers another point of interest is engineering's relationship to science, and national differences in the way this relationship is perceived. If we take Britain as a starting point, or for that matter the English-speaking world generally, there is a tendency to treat engineers as some kind of 'semi-detached scientists'. Engineering, indeed, is often labelled 'applied science', and engineers are thought of as having got the knowledge core of their subject, which forms the basis of their courses, from the natural sciences. They are bracketed with scientists in statistical breakdowns and in the QSE (qualified scientists and engineers) formula of the manpower planners.

Yet somehow or other there is the feeling that even if they are working alongside the scientists they will be doing this as helpers rather than equals, entrusted with hardware, implementation, and the overcoming of 'teething troubles', leaving the clever, conceptual stuff to their betters. The engineering part of the partnership is just a 'working-out-the-details' job and worrying about the practical trivia — the cost for example.

Our view is that this concept of engineering, as applied science, is misleading, and it is regrettable too that it conceivably contributes to the depreciation of engineering work as well as to its misapprehension. Here again, however, we have a phenomenon which is not necessarily trans-cultural. In Germany this particular misapprehension of engineering does not occur, the term 'applied science' is not current, and the notion is even rather difficult to formulate in German. We will explore this issue more fully in a later chapter where the factors relating to the engineer's standing in West Germany are discussed in detail. It is, however, an unusual feature of engineering that its substance may, in Britain at any rate, be defined in relation to another set of disciplines, the natural sciences, and the standing of its practitioners similarly relativized. Most occupations are, after all, defined in their own terms not in relation to a putative academic big brother.

Another unusual feature of the engineer's position is the variable degree to which his personal title, engineer, may be diluted. In Britain it is a common complaint that not only is the title 'engineer' not legally protected, it is also widely abrogated. Thus a wide range of maintenance fitters, machinists, repairmen, and semi-skilled operators call themselves engineers, are so labelled by others, and belong to trade unions having the words engineer or engineering in the title. This does not please British engineers, and is sometimes cited as a determinant (or reflection) of the engineer's relatively poor image and status in Britain. Now this phenomenon is not (completely) unknown in West Germany, but it is appreciably rarer, and German language conventions do not encourage this dilution. In German, that is to say, engineer-words are typically used in a precise qualificational way. *Ingenieur* is thus used to connote the holder of the Ing. Grad. qualification (a substantial non-graduate qualification discussed particularly in the next two chapters) while a person with a university degree in engineering generally describes himself, and is described by others, as a *Diplom-Ingenieur* (literally graduate engineer). Another ramification of this phenomenon is that in the German scheme of things there is a clear cut division between the non-graduate engineer and the technician. A distinction, that is, in terms of qualifications where the technician holds a quite different qualification from the *Ingenieur* (Ing. Grad.) obtained by means of a less demanding course at a different institution. The technician's (normal) job and promotional prospects are also circumscribed as opposed to those of the engineer. All this represents a contrast to Britain where there is no clear distinction between the technician on the one hand and the non-graduate engineer on the other. The non-graduate qualifications in Britain are also sometimes regarded as technician qualifications anyway, and this trend has been reinforced by the engineering institutions, in the sense of the institutions of the CEI, having a membership-grade system in which the holders of non-graduate qualifications are termed *Technician* Engineers whereas the superior status of Chartered Engineer is now reserved for graduates.

Yet another way in which the engineer is at the point of controversy concerns the issue of the *professional* standing of his occupation. In the English-speaking world some occupations are viewed as 'professions', and the label is very prestige conferring. Understandably occupational groups outside the traditional professions are keen to adopt the professions label, while the traditional professions may be equally keen to 'repel boarders'. Among the aspirant groups in Britain are, of course, the engineers who would like to be socially esteemed alongside doctors and lawyers. This state of affairs is interesting in two ways.

Firstly, and it is a composite theme, the engineer's claim to be classed as a member of a profession is quite genuinely controversial. It can be argued either way. On the plus side the engineer can claim a high educational level, a lengthy training, and the possession of specialized knowledge. To this may be added the twin claims that the practice of engineering is a responsible job, and one marked by a high level of social utility. All these considerations tend to place the engineer with those enjoying membership of the traditional professions, but this is not the complete picture.

For the engineer self-regulation by a professional peer group is weaker than is the case with the traditional professions (cf. the role of the Law Society for solicitors or the BMA for doctors). This is a reflection of the fact that most engineers are not 'independent professionals' after the manner of, say, lawyers. Engineers are usually employees, and they tend to work for large organizations typically manufacturing companies. This last fact has a further implication; it means that they are not guided by criteria of specialist technical excellence alone (after the ideal type model of the professional), but are subject also to economic and managerial constraints. Furthermore, being an engineer is not typically a terminal career state. Most engineers work in industry and aim to become managers. British engineers, certainly, are pleased to be able to describe themselves as managers.[5] There is a real contrast here with the traditional professions. A hospital doctor, for instance, aims to become a consultant (super doctor) not a medical administrator.

These points have some *sui generis* interest as a partial

characterization of engineering as an occupation. But more than this they may help to show what a distinctive occupational status engineering is. The second general point to emerge from this contentious issue of the engineer's professional standing is that this issue is only really poignant in Britain. In Germany, although the same set of arguments for and against can be voiced, the issue is not in fact canvassed, has little significance for the German engineer's social prestige, and is difficult to formulate in the German language.

The British concern over the engineer's professional status may in general terms be seen as compensatory. In particular, the ascription of professional standing to our engineers would compensate for the largely rather unflattering stereotypes of which they are the victim. This unflattering stereotyping of engineers is not exclusively British, though it is fair to say that it is probably more marked in Britain than elsewhere. Neither are engineers the only occupational group to be stereotyped — accountants, vicars, school teachers, and salesmen are also widely stereotyped, not to mention professors, colonels, and income tax inspectors. Yet the stereotyping of engineers, especially in Britain, does embrace some unusual elements.

First, this stereotyping is based on very slight evidence. Whether or not our stereotypes of say vicars and school teachers are valid, we have all, at any rate, had opportunities to observe members of these groups and what they do is generally known. Engineers on the other hand have rather little public exposure. Non-engineers have in fact little idea of what it is that engineers do all day. So any stereotyping is not so much a distortion of reality as a substitution for it. Second, it is odd that the stereotype is at least slighting if not frankly hostile. This is unusual in the stereotyping of occupational groups; one has to turn to the stereotyping of nationalities to find comparisons. And third, the stereotype contains some oddly juxtaposed elements. That the 'typical engineer' is, for instance, assumed to be lacking in personal refinement, a 'rude mechanical', probably rather boring and not the sort of chap to enliven a soirée. Then there is the view that engineers have a common political character, or at any rate fit on adjacent parts of the continuum.

They are thought to be right wing, uninterestingly conven-

tional in their socio-political views, and submissive to authority to boot; and if not this then at least culpably apolitical. And
lastly there is pure occupational demotion whereby the
engineer is characterized as a fitter, with allusions to oil cans
and dirty hands. An odd aspect of all this is that these three
elements are not interconnected: they do not imply, cause, or
condition each other. So the stereotype is not only unflattering
and literally unfounded — it is also a logical jumble in which
elements with as little in common as wife-beating and hay fever
are juxtaposed. It is also the case that thanks to an extensive
German study[6] it is now possible substantially to repudiate
major elements of the stereotype, and this is one of the themes
explored in the pages which follow. Furthermore, this explo-
ding of the stereotype in the German case should be viewed as
applying to engineers in general, unless there is any counter
evidence — and there is not.

The relatively higher standing of the engineer in West
Germany is, as was suggested earlier, a theme and an organiz-
ing principle of part of this book. It can also be claimed that a
related theme is explored, namely that of the status of industry.
This expression has been current, indeed fashionable, in
Britain in the last few years, together with the corollary that the
status of industry here is not high, and the implication that
industry enjoys more standing in some other countries. Using
primarily survey data on German engineers we are able to
make a threefold contribution to this status of industry idea.

First, it is possible to give some tangible idea of what it
means in practice to say that the status of industry is higher in a
given country. Second, at the same time we can offer some
evidence in support of the claim that industry does enjoy a
more favoured position in West Germany. And lastly we are
able to connect the ideas and the facts of the status of industry
on the one hand and the standing of the engineer on the other.
It is always gratifying to close circles.

Finally in these introductory remarks we might extend the
opening idea, that of the possible differences between industrial societies. If these differences are real they can only be
mapped on the basis of knowledge, and to this end we aim to
furnish the results of our own and other studies. This is a
humble and straightforward aim, but one which should not be

lost in assumptions about either homogeneity of European countries or the plenitude of existing data. For some time people have believed that they know, or get told quickly, what is happening in other parts of the world. This is largely true, but it is a question of being informed about incidents rather than institutions. Or, to put it another way, the idea of the 'global village' refers to information not knowledge, and the presence of the former may obscure the lack of the latter. We hope to dispel a little of the obscurity.

Chapter 2

Education and Training

The recruitment and training of engineers are closely related themes. In the present chapter we will offer some account of the German education system, thus making clear the actual training processes and showing the educational context within which the recruitment of engineers occurs in Germany. The following chapter will then discuss against this background the recruitment of the engineers in the survey.

It would be possible to offer a briefer account of the German education system than that presented here, by singling out those parts of the system which refer explicitly to the education of engineers and discounting the rest. However, our view is that to treat higher technical education out of context would be misleading, and in any case the German education system is interesting in some general ways, several of which in turn are relevant to the training of engineers.

First, the German system offers, or has offered, a number of alternatives. As will be shown in the following account there are different kinds of *Gymnasium* (the German equivalent of the English grammar school), different ways of entering university to read for an engineering degree, and in some cases different institutions at which the same engineering qualification may be obtained.

Second, these differences and alternatives within the system are similarly enhanced by the fact that West Germany is a federal republic and the *Länder* or states making up the federal republic each have the right to determine their own educational system. This right to cultural sovereignty, as it is called, gives rise to differences in, for instance, the terms on which nursery education is available, the presence or absence of comprehensive schools, the extent of practical training in the courses for non-graduate engineers, and so on. These two features, alternatives within the system and variations from state to state, are also worth mentioning since they run counter to the image some people have of Germany as a more standardized and bureaucratic society than Britain: there is no evidence for such an image in terms of the educational system as will be demonstrated in the following account.

Another point of interest, and it is discussed in some detail in this chapter, is the extent of the German provision for vocational education, again characterized by the range of alternatives and options. Germany is a society which takes vocational education seriously, a theme which is developed not only here but also in a subsequent chapter on the status of industry. In this there is some contrast with Britain, and an even sharper contrast with France.

This chapter also includes a quite detailed discussion of a recently created institution of further and higher education, the *Gesamthochschule*. In a way this is not justified because only a relatively small proportion of German engineers are trained at the *Gesamthochschule*; on the other hand the *Gesamthochschule* embodies several features of the German system and ethos, and has an almost symbolic status. It exemplifies *par excellence* the inherent complexity of German educational arrangements; it is also a manifestation of the German propensity to experiment in education, and shows clearly the role of ideals in educational change in present-day Germany. Furthermore, some critics would argue that it illustrates a German capacity to mess up a very sensible system, albeit for high-minded reasons.

The educational routes taken by aspiring engineers begin of course at school, and so does our account of the education system. This system is complicated, but we try here to reduce it to manageable proportions.[1]

It should be noted that in Germany, as in most other European countries, the educational system has been changing during the last two decades, owing to social and political pressures. In 1964 the school leaving age was raised to fifteen, since when there has been a development towards a more flexible and socially equitable system of secondary schools providing the possibility of alternative routes to university. We are therefore viewing a shifting scene and, at the time of writing, find it most convenient to divide the German educational system into four main stages as follows:

1. Pre-school   (under 6 years)

2. Primary   (6–9 years. Classes 1–4)
3. Secondary Stage 1 General academic (10–15 or 18.
              Classes 5–10 or 13)
              Stage 2 Vocational (16–18. Classes 11–12)
4. Tertiary   Universities and colleges (19 and over).

The Federal Government shares with the *Länder* (States) the tasks of planning and research at a national level, but each *Land* (State) is responsible for its own education, administration, and inspection. A standing conference of the Ministers of Education attempts to maintain reasonable uniformity throughout Germany.

Compulsory education begins at the age of six in the *Grundschule* (primary school) where children stay four years until, at the age of ten they move to one of the three types of state secondary school. Those suitable for university spend nine years at a *Gymnasium* (grammar school, ten to nineteen). The next academic stream spend six years at a *Realschule* (middle school), and the remainder spend five or six years at a *Hauptschule* (secondary modern). There are very few private secondary schools, mainly girls' grammar schools, commercial schools, and clerical colleges, but even these derive 90 per cent of their income from public funds. It should be added that these private schools should not be regarded as equivalent to the public schools in Britain; no prestige attaches to attending a private school in Germany, and they are not regarded as educating an élite.

For those who cease full-time general education at fifteen, further part-time education is compulsory from sixteen to eighteen. Thus all children have compulsory education of some kind from their sixth to eighteenth year and those in the *Gymnasium* (grammar school) stream have one year longer, leaving school at nineteen.

Most pupils attend so called half-day schools, which typically operate somewhere between about 7.30 a.m. and 2.30 p.m. on weekdays and also on Saturdays. About one quarter of our engineers went to 'full day' schools.

PRE-SCHOOL EDUCATION

Pre-school education is available but not compulsory and is financed mainly on a private basis. There are voluntary

Kindergärten (kindergarten) for three-to-six-year-olds which get 75 per cent of their funds from private sources such as churches, business firms, donations, and bequests from individuals.

For school-age children who are so physically or mentally handicapped that they are not ready to take their place in the main school system there are *Schulkindergärten* (kindergarten for handicapped) established and financed by public authorities. Attendance is free and is voluntary, except in Bremen.

PRIMARY EDUCATION

The primary school for six-to-nine-year-olds is called *Grundschule* and the *Länder* (states) have each developed *Lehrpläne* (teaching programmes) to serve as a guide for the course content and methods of instruction.

As in England there is a trend away from a particular teacher for each class towards subject teachers. For the first two or three years, two thirds of the pupils' time is spent with class teachers, whereas in the third and fourth years it is only about two hours daily.

At the age of ten, children move on to secondary schools. The type of secondary school that the child will attend is chosen by the parents with the advice of the primary school teachers, but the pupil's progress during the first two years is monitored and the *Land* (state) reserves the right to transfer any child to a more suitable school if it is not progressing satisfactorily.

SECONDARY EDUCATION: STAGE 1

There are three main types of secondary school. The *Hauptschule* roughly equivalent to the former English secondary modern school,[2] taking 45 per cent of the children in the age group (ten to fifteen); the *Realschule,* formerly called Middle School, and of an intermediate level, higher academically than *Hauptschule* but still vocationally oriented and taking 20 per cent of children; and the *Gymnasium,* the highest level, academically equivalent to the English grammar school taking 25 per cent of children; with the remaining 5 per cent attending special schools. It is noteworthy that, compared with England, there has been little development of comprehensive schools. Despite social and political pressures less than 10 per cent of the secondary school population are in comprehensive schools and most of these are in Hessen or West Berlin, which are the most left-wing states. It seems therefore that Germans, irrespective of political ideals, are more ready to accept a meritocratic educational system provided the opportunities are there for all.

Entry to *Realschule* and *Gymnasium* is by a combination of parental choice and recommendation from *Grundschule* with the additional provision in some *Länder* of 'trial instruction'. There is a probationary period for the pupil of up to two years, during which reallocation may occur if the *Land* thinks this is desireable.

Hauptschule

This type of school means, literally, main school; previously it was called *Volksschule* (people's school). Such schools provide education from ten to fifteen, or sixteen in some *Länder,* with a view to vocational training. At fifteen or sixteen pupils can choose to go on to full or part-time vocational courses in various types of technical and commercial colleges which will be described later.

Teaching in a *Hauptschule* is mainly on a subject-teacher basis and includes instruction in handicraft subjects and one foreign language.

Pupils who complete the *Hauptschule* course satisfactorily obtain an *Abschlusszeugnis* (school leaving certificate). In most *Länder* this is awarded on continuous assessment without the pupils passing a final examination. Some *Länder* award an additional certificate, the *Qualifizierter Abschlusszeugnis* (qualifying leaving certificate) which admits pupils to Stage 2 vocational secondary schools with courses leading to the *Fachhochschulreife*.

It is also possible for the best final year pupils to transfer and continue with a grammar school education. For this purpose there are special grammar schools called *Aufbaugymnasium* to provide the conversion.

Realschule

There is no direct English equivalent to this type of school, it was previously called *Mittelschule,* literally middle school, a term which describes it very closely. The courses provided are intermediate in level and type between the *Hauptschule* and *Gymnasium.*

Entry to *Realschule* is by parental choice and recommendation from *Grundschule.* Pupils (ten to fifteen) follow a six-year course and obtain an *Abschlusszeugnis der Realschule* (leaving certificate of *Realschule*). This certificate enables the pupils to enter a wider range of Stage 2 secondary schools than from *Hauptschule.* As with the *Hauptschule* best pupils it is also possible for the best leaving *Realschule* to enter an *Aufbaugymnasium* offering conversion courses of an academic type which enable them to transfer into the university stream if good enough.

Gymnasium

At this type of school pupils follow a nine-year course (ten to nineteen) leading to the qualification of *Abitur,* which is equivalent to the English GCE A-level and in principle gives right of entry to university. The *Gymnasium* therefore embraces both Stage 1 and Stage 2 of secondary education in Germany.

There are three main types of *Gymnasium,* those for classics, those for modern languages and those for mathematics and science. In addition there are some *Gymnasia* specializing in subjects such as music, domestic science, agriculture, and physical education.

Formerly entrance was by examination, but it is now based on assessment and recommendation by *Grundschule* and by parent's choice. If at the end of two years at *Gymnasium* pupils are unable to cope with the courses, they can be transferred by the *Land* to a school more suited to their ability.

The *Abitur* is based on the last three years of study at *Gymnasium* in which five main subjects are taken, together with four subsidiary ones. The latter, which only extend over the first two years of the *Abitur* course are non-examinable, comprise religion and physical education which are obligatory, and two optional subjects. It will be noted that the *Abitur* examines a wider course than the English A-level examination.

Each of the five main subjects involves a four-hour written examination. Although the range is wider than in GCE A-level, both mathematics and German are obligatory. For instance in a classics *Gymnasium,* mathematics and German must be two of the subjects, and in a mathematics–science *Gymnasium* students must take German and one foreign language.

Written examinations are set and marked by individual schools and moderated by the *Land*. All subjects ever taken at *Gymnasium* are listed on the *Abitur* certificate and for the main subjects grades 1–6 are awarded. Grade 1 is outstanding, 5 is a failure, and 6 a bad failure. Students with say two 5s or one 6, must repeat the whole year. Because of this and the broad nature of the examination it is not easy to maintain uniformity and there is a Federal Government department in Bonn which compares standards. Nevertheless it seems to be accepted that standards vary. For example, because Bavaria tends to overgrade and Hessen tends to undergrade, due allowance should be made when comparing candidates from such *Länder*.

By the 1949 *Grundgesetz* (basic law), the *Abitur* gives the right to attend university and to read any course. In practice, however, and particularly where there is a selective entry (*numerus clausus*), because of the excessive demand for places in subjects such as medicine, students cannot automatically enter any university to read what they want (see page 21).

SECONDARY EDUCATION: STAGE 2

Those pupils who do not continue their Stage 1 secondary school beyond their fifteenth year, that is after ten years schooling, must still attend one of the many types of Stage 2 secondary school for at least another two years. These schools, listed below, provide either full-time or part-time vocational or craft training and cover a wide range of courses. The approximate percentages of the age group (including the grammar school stream — 13 per cent) for the year 1970 are given in brackets:

Berufsschule — part-time
Berufsgrundschule — full-time   } (70)
Berufsfachschule — full-time   (9)

Fachoberschule — full-time or part-time   (4)
Berufsaufbauschule — part-time,
   full-time, or evening courses   (1.5)
Fachschule — usually full-time   (4)
Aufbaugymnasium   —

BERUFSSCHULE

Attendance is obligatory for those who have completed their
ten years of full-time education, but who are not attending any
other kind of academic or vocational school, for those who
have not reached the age of sixteen, and for those who have not
completed their apprenticeship and are still undergoing
training.

The courses give training for craft work, jobs in commerce,
business, or administration, and junior to middle-grade posts
in public service. They are usually for three years on a part-
time 'day-release' system of eight to twelve hours per week or,
less often, block release.

BERUFSGRUNDSCHULE

This is a new type of school for those who have had ten to
eleven years full-time education, the last six years at either
Realschule or *Hauptschule*. It provides full-time courses with
the aim of helping students to choose an occupation or occupational group, after one year's instruction.

BERUFSFACHSCHULE

This school provides a full-time one- to three-year course of
vocational training in such fields as commerce, trade, hand-
crafts, or social work. For some jobs such as technical assistant
it provides a complete training.

Entry requirements are at least the *Hauptschulabschluss* or
the *Abschlusszeugnis der Realschule,* for neither of which are
any formal examinations required. In some schools after two
years of instruction students leave at the standard of
Fachoberschulreife (the level for entry to *Fachoberschule*).

FACHOBERSCHULE

For pupils from a Realschule who wish to progress to higher
education, this type of school provides a further two or more
years of vocationally oriented education.

The curriculum includes general education, subjects related to the chosen vocational field of study and related practical training during the first year. They may thus be compared with colleges of further education in England.

The entry requirement is usually the *Abschlusszeugnis der Realschule.* Courses are two-year, or, more rarely, three-year part-time or full-time, leading to the award of *Fachhoch-schulreife* which gives entry to *Fachhochschulen* (engineering schools) and thence possibly to a technical university.

BERUFSAUFBAUSCHULE

This type of school is being replaced by the *Fachoberschulen* described above. It offers a general and vocational training beyond that of *Berufsschule,* either part-time or full-time.

Such a school provides a link between the general and vocational educational systems. Instruction in a full-time course is for two to three half years and in a part-time course ten to twelve hours a week for a period of six or seven half-years. Various mixes of full- and part-time courses are available.

The leaving qualification is the *Fachhochschulreife* which gives entry to the specialist types of *Gymnasium* already referred to (music, domestic sciences, etc.) or to the *Fachhochschule* in the tertiary sector of education.

FACHSCHULE

These schools are equivalent to technical colleges (now colleges of further education) colleges of commerce or trade schools in England, and provide further training for most occupational sectors, and include:

Technician schools
Chief-technician schools for craftsmen and industrial trades
Agricultural schools
Schools for business economics, administration, mining, domestic science, rehabilitation of the handicapped, and for textile and clothing manufacture.

Entry requirements are the *Abschlusszeugnis* from *Hauptschule,* a practical ability in a vocation, and a technical-vocational education.

The courses for full-time students last at least six months, but are more often one to two years. For many occupations the vocational training can be completed by taking the *Technikerprüfung* (final state examination) which represents the technicians' accolade.

TERTIARY EDUCATION

So far we have seen that German secondary education which is freely available to all involves either a grammar school type of education, or a shorter period of general academic education followed by some kind of full- or part-time vocational or technical training. For successful students both streams can nowadays lead to the tertiary system which comprises:

1. Universitäten (universities)
2. Technische Hochschulen (technical universities)
3. Pädagogische Hochschulen (colleges of education)
4. Kunst /Musik /Sport Hochschulen (colleges of art /music / sport)
5. Fachhochschulen (engineering schools)
6. Gesamthochschulen (comprehensive universities).

Although academic and registration fees are nominal, the subsistence and maintenance fees for a student are considerable. This tends to encourage students to attend their local university and to make the tertiary system more the province of students from middle- and upper-class families who can afford to finance them through the six years of university.

The two main routes to an engineering qualification are the *Technische Hochschulen* (technical universities)[3] and the three general universities, Bochum, Erlangen, and Trier-Kaiserslautern, having an engineering faculty leading to the *Diplom Ingenieur* (Dipl. Ing.) qualification, and *Fachhochschulen* (engineering schools) leading to the *Ingenieur Graduiert* (Ing. Grad.) qualification. The system is, however, broadening, and it is now possible to obtain a qualification called Dipl. Ing. at a *Gesamthochschule* (comprehensive university), but at the time of the survey no students had reached graduation by this route.

In Germany the two professional engineering qualifications are awarded by the state and have been protected by a federal

law since 1967. The Dipl. Ing. is recognized as the fully qualified professional engineer and the Ing. Grad. as a somewhat
lower status but nevertheless qualified engineer. It seems
unavoidable that the new Dipl. Ing. from a *Gesamthochschule*
will be lower in status than the conventional Dipl. Ing., as will
be made clear later.

The route chosen depends largely on the type of secondary
school attended and the educational level reached. Most of
those who went to *Gymnasium* and obtained *Abitur* go to one
of the twelve technical universities; whereas those who were at
Realschule, or *Hauptschule* and left school at the age of sixteen
with the *Fachhochschulreife* go to a *Fachhochschule* via either
a *Fachoberschule* or *Fachschule.* Having obtained the Ing.
Grad. qualification, it is not unusual for students to go on to
university to take the Dipl. Ing., 6 per cent of Dipl. Ing. have
done so. Of the weighted sample surveyed 55 per cent qualified
at *Ingenieurschulen,*[4] 18 per cent at *Fachhochschulen,* 16 per
cent at *Technische Hochschulen* and 7 per cent at *Technische
Universitäten.* The new third possible route will be discussed in
detail later.

TECHNISCHE HOCHSCHULEN AND UNIVERSITÄTEN

Although the universities in Germany were founded relatively
late compared with those in the rest of Europe, the origins of
the technical universities were relatively early. In 1745 the
Carolo-Wilhelmina Collegium was founded in Braunschweig,
followed by the *Bergakademie* (mining school) at Clausthal
Zellerfeld in 1775 and a *Bauakademie* (building school) in
Berlin in 1799. In Britain the first universities came early by
comparison and the engineering schools late.

Now there are about forty general universities in Germany,
only three of which, all modern, have a faculty of engineering
which award the Dipl. Ing. The same qualification is more
commonly obtained at the twelve older established technical
universities.[5]

The Dipl. Ing. course takes a minimum of nine semesters
(4½ years) but in practice the average time taken is 5.8 years.
Although the teaching staff are usually experienced engineers
from industry, the course is more academic in nature and goes
deeper into fundamentals than the Ing. Grad. course at the

Fachhochschulen. Nevertheless the German university course is longer and more practically oriented than engineering courses at English universities.

STAFF STRUCTURE

In German universities all professors are appointed by invitation after careful vetting and selection compared with Britain where candidates may apply in response to advertisements. Moreover most German professors of engineering come from a senior post in industry and maintain close contacts with industry thereafter. In Britain very few engineering professors are appointed direct from industry but have previously spent several years in university as a senior lecturer or reader. Their background is therefore academic and research in contrast to the German tradition of industry and design.

In the *Fachhochschulen* and *Gesamthochschulen* nowadays the new teaching staff must have university degrees and must have spent at least five years in industry. At both university and engineering school level there is therefore greater insistence on industrial experience and ability for teaching staff.

HIGHER DEGREES

A higher proportion of German engineering graduates go on to become Dr. Ing., roughly equivalent to the British Ph.D. in engineering. One tenth of German engineers (and also the same proportion in USA) have a doctors degree, whereas in Britain only about one in twenty-five graduates has.

PRACTICAL TRAINING FOR THE DIPL. ING.

Up to the 1960s all university engineering students had to complete fifty-two weeks of industrial training before they graduated; half of this before entry to university. However since then the total period has been reduced to twenty-six weeks, again with half (i.e. thirteen weeks) before entering university. This is very much regretted by university staff and industry. The change was because the *Abitur* examinations which originally were at Easter, allowing six months before entry to university, are now held in the summer and leave only three months for preliminary training. At this stage *Abiturients* are already nineteen compared with the Ing. Grad. students who start practical training at about eighteen and graduate by twenty-one.

Each student must find a practical training place in industry for himself. The universities do not usually help with finding places or keeping an eye on the trainee during this period. Their function is to draw up guide lines for the conduct of the practical training and how it is to be recorded and reported by the student. These requirements which are universal and similar throughout the universities are regulated, along with curricula, at the annual meetings of the *Fakultätentag Maschinenbau* (standing conference of mechanical engineering professors). Other branches of engineering are regulated by similar standing conferences appropriate to the subject. Each university has a junior member of staff responsible for examining the log-books and reports and checking with the student that he knows about the work that he has described. Although the universities do not therefore have any direct control, the practical training system seems to work well, not only in large firms but also medium-sized ones. This is perhaps because the trainees are usually under the supervision of an older and senior engineer in each company, who has himself been through the same system.

Most students obtain a place in their home area or where they are going to study. Trainees (usually twenty years old or even twenty-two, if they have already done their military service) are paid about the same as a second-year apprentice of sixteen. The type of industrial training is therefore linked with routes through secondary and tertiary education.

SELECTIVE ENTRY TO UNIVERSITY

In many technical universities there is now a limitation on numbers (*numerus clausus*) and a selective entry in the more popular subjects. Thus it no longer means that the *Abitur* holders can automatically enter the university of their choice and study what they want. This is common in medicine, economics, and philosophy, but to date has not greatly affected engineering students owing to their relative scarcity.

For subjects in which there is a *numerus clausus,* applications are handled by a central office at Dortmund called ZVS (*Zentralstelle für die Vergabe von Studienplätzen*) which was set up in 1972. Selection is based on average *Abitur* grades, the time during which the applicant has waited since obtaining his *Abitur* and any individual hardship. The central office goes further than the English UCCA and itself decides on the university to which the student must go, irrespective of any preference that the student may have.

This system is naturally unpopular among students, and also the technical universities, several of which have flouted the system and challenged the ZVS office on its constitutional authority to make such decisions. It seems therefore that a piecemeal use of the *numerus clausus* and the ZVS have become political issues but students still find ways of evading the system.

FACHHOCHSCHULEN

Until 1968 the *Fachhochschulen* were termed *Ingenieur-schulen* and provided a three year full-time course for the *Ingenieur* (Ing.) qualification, which represented something like the English Higher National Diploma in engineering or an ordinary degree. Now there has been a further change in that many *Fachhochschulen*, particularly in *Nordrhein Westfalen*, have been incorporated in the new *Gesamthochschulen* (see below).

To enter a *Fachhochshchule* students must have attained the *Fachhochschulreife* from their secondary education and have completed several weeks' industrial training. The *Fachhoch-schulreife* does now admit a limited number of candidates to university, although previously this had been exclusively with the *Abitur* qualification; and conversely the *Abitur* admits a candidate to a *Fachhochschule*, but few choose this route.

Courses last for six semesters (usually three years) and lead to the Ing. Grad. qualification. Originally the Ing. qualification meant that the holder had received not only a good three year technical education, but also an initial very thorough two year craft training. This is no longer the case with the Ing. Grad., because the course is more academic than previously and the practical training is shorter and at a later stage. There is a danger therefore that the good 'practical engineer' will become 'extinct'. The consequences of this development are discussed in a very interesting book.[6]

PRACTICAL TRAINING FOR THE ING. GRAD.

Formerly the non-university student did two years practical training immediately after attaining the level of *Mittlerreife* at *Mittelschule* (now *Realschule*) at the age of 16. But since the *Ingenieurschulen* were changed to *Fachhochschulen* in 1968, students must go to a *Fachoberschule* and spend two years on a mixed general and technical course before entering a *Fachhochschule*. In the two years at *Fachoberschule* students may commence a formal apprenticeship, although it is not now essential.

In Germany a craft apprenticeship normally lasts three and a half years, but it may be completed in a shorter period if the apprentice has gained the equivalent of the English O or A-level GCE.

It was seen in the survey that 71 per cent of the sample (84 per cent Ing. Grad., 24 per cent Dipl. Ing.) had served a formal apprenticeship. This appears to be a remarkably high proportion but it must be remembered that most of those with the Ing. Grad. qualification would have attained their degrees under the old system of *Ingenieurschulen*. But it also seems that a significant number with *Abitur* chose to serve an industrial apprenticeship rather than go straight to university.

Practical training arrangements at *Fachhochschule* are complicated, they vary from subject to subject, *Fachhochschule* to *Fachhochschule* and from *Land* to *Land*. Training may be before or during the course or both.[7]

Some 94 per cent of the sample in our study believed that practical training was necessary and the majority thought that it was best done before commencing academic studies; in fact 91 per cent had done their training then.

STUDENT GRANTS

Since the so called *Bad Honnef* agreement in 1955, it has been possible for students to obtain what is largely a loan to help them through university (and *Fachhochschule* or *Gesamthochschule*). This has since been replaced by a federal law with the short name of *Bafög* (*Bundesausbildungs-förderungsgesetz*).

To obtain such loans students have to show financial hardship and strong family financial reasons for requiring assistance. At the moment less than one quarter of students have loans; and even a full loan does not cover more than two thirds of the cost. For example the income of skilled manual workers is usually too high for such families to qualify for a loan unless there are many children in the family, and as a result the proportion of university students from working-class families is only about one-eighth in Germany compared with one-quarter in Britain.

Before 1955 and during and before the Second World War, the proportion getting grants and loans from church and private foundations was extremely small and it can therefore be said that despite a state school system those going on to university tend to be those who can afford to pay.

STUDENT WASTAGE

In making comparisons with England it should be pointed out that the loss rate of students in engineering courses in West Germany is relatively high and that in fact about one third leave after one year, rising to about half, if losses in later years are included. This is in contrast to England where in engineering faculties the over-all loss rate is about one in ten.

GESAMTHOCHSCHULEN

Since there is no precise equivalent in the English system of the German *Gesamthochschulen,* a recent educational development, it was felt that a few notes on their structure and aims would be useful, particularly if there were a move in England to alter the training of engineers or to try to change the image of the profession by imitating the German system. The following observations are based on visits to several *Gesamthochschulen,* official brochures, reports etc.

The word *Gesamthochschule* means literally 'comprehensive high school' with the high school having a higher education connotation, not the secondary school connotation as in the United States. The basic idea is that a variety of higher and further educational establishments in a town or area, should be amalgamated to form one *Gesamthochschule,* which would offer a complete or wide range of courses at university level and a considerable range of courses, mostly vocational, at the sub-university level, roughly equivalent to HND in England. It should be pointed out that this level of qualification in Germany exists for a wide range of subjects, mostly vocational and not just those which are technical; for instance one can gain a qualification in economics as well as engineering (Ing. Grad.).

As well as integrating institutions the aim is to integrate course types and levels by means of so called Y model, H model, and straight-line courses (see below for details). It is also to integrate the teaching body by taking on new staff from the traditional university sector (referred to from now on as university teachers), and amalgamating them with the staff from the various educational institutions which are being combined under the one *Gesamthochschule* (these will be referred to as the technical college teachers).

For political reasons the *Gesamthochschule* movement is primarily a *Nordrhein-Westfalen* development and up to 1977 five had been founded. Those being developed in other *Länder* are not regarded as 'real' *Gesamthochschulen* by *Nordrhein-Westfalen* standards, because of the limited scope of their integrative activities, the limited range of institutions involved in the amalgamation, or the narrow range of courses offered. All the *Gesamthochschulen* visited were in *Nordrhein-Westfalen* or *Hessen*.

The educational institutions making up any particular *Gesamthochschule* will vary from place to place. Generally it will consist of one or more *Fachhochschule,* a *Handelschule* (college of commerce) and a *Pädagogisches Institut* (teachers training college), with the staff and facilities for university level courses being added. One *Gesamthochschule* visited comprised a medical clinic with university status, a *Fachhochschule,* and a *Pädagogisches Institut,* plus new university components; another was the result of an amalgamation of forty former educational institutes, including two *Fachschulen,* two *Ingenieurschulen* (one for architecture and civil engineering, the other for electrical and mechanical engineering), *Handelschule* and a *Kunst und Musikschule* (art and music college). It is not unusual to find that a *Gesamthochschule* is on a split site or is multi-site especially in the first years. Where the will and resources are available, however, there may be an extensive building programme aimed at bringing the whole organization onto one central site; the *Gesamthochschule* at Essen is an example of this.

Before remarking on points of weakness and operating difficulties in the *Gesamthochschulen* we would first make it understood that these should not be interpreted as a general criticism of the movement (which would be premature in any case), but rather to bring an awareness of the difficulties that can arise and which will have to be solved for the movement to be successful. Secondly we would warn against a kind of cultural short circuit. When considering the institutions of another country it is tempting to assume that these are more or less like those of one's own country, and all that is required is to spend a few minutes at the beginning working out the system of equivalents. In this case one might assume that the German *Gesamthochschule* is more or less the same as the English polytechnic, college of higher education, or college of further education. There is a certain superficial justification for this in that our polytechnics also have status problems, and our colleges of higher education and of further education are also massively heterogeneous, offering a wide range of courses. It may therefore be helpful to note a few ways in which the German *Gesamthochschulen* differ from any English educational institution.

1. The *Gesamthochschule* movement is much more firmly entrenched in an ideology than are the English polytechnics. There is more of a tendency for colleges of art and music to combine with technical colleges, although this is now beginning to happen in England, but for economic rather than ideological reasons.
2. The English institutions may house courses of different kinds and levels, but they do not try to integrate them. For instance a polytechnic would not combine an HND with a degree course by means of Y-model courses.
3. The heterogeneity of courses in English institutions has developed piecemeal and gradually, reflecting local demand and local initiative; whereas in a German *Gesamthochschule* this heterogeneity has been produced 'at a stroke' when the amalgamations were effected.
4. In the *Gesamthochschulen* there is a sharp cleavage between the *Fachhochschule* teachers and the university

teachers (see below). Any staff cleavage in the English institutions has been less acute, and indeed it is less easy to categorize the staff of an English polytechnic as 'technical college lecturers' or 'university lecturers'.

AIMS AND IDEOLOGY OF THE GESAMTHOCHSCHULEN

The following remarks made by one of the *Rektors* during the visits to *Gesamthochschulen* seem to sum up their aims:

The Gesamthochschule is not an exercise in leveling. The purpose is to promote learning in a unified institutional setting, in which teaching tasks are shared. There has been criticism that the *Fachhochschule* students have been educated in a way that was too much like school. Similarly there has been criticism that the universities have been too 'ivory tower' like. To give the two types of students the best of both worlds, intellectual flexibility and practicality is the mission of the *Gesamthochschulen*.

However they do omit reference to the ideological side of the development.

It is not an accident that the stronghold of the *Gesamthochschule* movement is in the traditionally *Socialistische Partei Deutschlands* dominated *Land* of *Nordrhein-Westfalen*. The ideological case is a dislike of status differences in principle, including those resulting from, or reinforced by, attendance at educational institutes of different status which confer qualifications of different status. Associated with this is a strong desire for equality of opportunity, and the removal of traditional competitive pressures. This means, in practice, freedom for students who had not attended *Gymnasium* and who do not have the *Abitur* qualification to pursue a degree course; on the other hand the students who have had the educational background which would traditionally have taken them on to a university degree are not obliged by the *Gesamthochschulen* course models to follow a conventional course to degree standard, if they lack the ability or inclination. In either case, the formal distinction between the two types of qualification is removed by standardising the title, but in practice there will still be social distinctions.

As the *Rektor's* remarks quoted suggest, the aim of the *Gesamthochschulen* is to avoid the disadvantages of the old type institutions, i.e. the limitations and school-like tendencies of the *Fachhochschulen* and the 'ivory tower' proclivities of traditional universities, and to give both types of student the best of both worlds. That is, that the *Fachhochshule* student should be exposed to a stimulating academic environment and the university student should benefit from greater emphasis on practicality and vocational relevance.

THE OPERATION OF THE SYSTEM

The following notes describe the system as it affects engineering education, though they are still valid for all subjects in which there is a two-tier qualification from educational institutions other than *Gesamthochschulen,* that is qualifications corresponding to the Dipl. Ing. and Ing. Grad. in the engineering field.

Students at a *Gesamthochschule* come from two distinct educational backgrounds:

1. Those who have attended a *Realschule* or *Fachschule,* have the *Abschlusszeugnis der Realschule* and have attained the *Fachhochschulreife* at a *Fachoberschule.*
2. Those who attended a *Gymnasium* and attained the *Abitur* qualification giving a constitutional right to admission to a conventional university.

During the visits to *Gesamthochschulen* it was never suggested that there were any major problems in accommodating the two types of students initially on the same course although some departments run a bridging course in key subjects for students without *Abitur.* These courses take place before the start of the first *semester* in the academic year and last for a few weeks.

The main course is usually organized in the form of a Y model.

The students from both educational backgrounds do a common two year course, which is followed by an intermediate examination. On the basis of the examination results, students are allocated to a further two year course (long course) or to a further one year (short course). In both cases the final award is called the Dipl. Ing. and one can only distinguish between the qualificational levels by referring to Dipl. Ing. (long course) or Dipl. Ing. (short course). A minority variation on this is the H

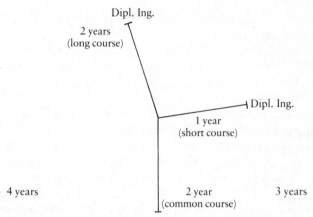

model in which the first two years are spent in parallel courses, following which the allocation to the long and short course is again made after an intermediate examination. One *Gesamthochschule* visited operated another variation, a straight-line three-year common course for all students, but this is probably an exception.

The Y-model system is typical for engineering and for other two-tier qualification subjects. Both the three- and four-year courses are flexible and students will often take longer to complete them. This *Gesamthochschule* situation reflects that in the *Fachhochschulen* and in the universities where the pre-scribed course lengths are three and four and a-half years respectively, but the average time actually taken by students to attain an Ing. Grad. or Dipl. Ing. qualification is 3.5 and 5.8 years respectively.

THE MERITS OF THE SYSTEM

It would seem that the aims and ideology of the *Gesamthoch-schulen* development as described above, are fully justified in principle, although in practice the difficulties of operating the system do inhibit the complete realization of the goals.

There is no doubt that for students without an *Abitur* quali-fication, a *Gesamthochschule* offers a real possibility of suc-cessfully completing a degree course. However it was pointed out by informants from both *Gesamthochschulen* and uni-versities that at present in a typical university department of mechanical engineering, about one third of the students would not have the *Abitur* qualification, but would have reached university by obtaining the Ing. Grad. qualification at a *Fachhochschule*. Under the *Gesamthochschule* system such students would proceed directly to the long course leading to a Dipl. Ing.

The law for establishing *Gesamthochschule* in *Land Hessen* was passed in 1970 and that for *Nordrhein-Westfalen* in 1971, but the actual setting up of the institutions came a little later. Therefore at the time of the survey (1976) it was not possible to find out how the greater flexibility was working because none of the *Gesamthochschulen* had completed course through-puts. Also it will be a few years before any judgement can be made on the attitude of employers to the new qualifications and how the graduates fare in the labour market.

For the members of staff at a new *Gesamthochschule* who formerly taught at a *Fachhochschule*, there are now possibilities to work for higher degrees if they wish, and to participate in research. In 1976 it was suggested that only a few had taken advantage of the new opportunities, but the opportunities are now there and may be taken up to a greater extent in the future.

A further merit of *Gesamthochschulen* is the great ease with which the contents of courses can be varied, especially to include subsidiary subjects from other departments. This is particularly valuable for engineers who may easily take options in subjects such as economics, sociology, and law. At one of the *Gesamthochschule* visited a member of the administration jocularly remarked 'our enemies say that a third of the engineering course is sociology'.

THE PRACTICAL PROBLEMS OF THE GESAMTHOCHSCHULEN

To complete this short survey of the new *Gesamthochschulen* system we must point out the considerable difficulties which have been encountered so far.

First the problem of a dual student intake, those with the *Abitur* qualification and those without. Most commentators felt this could be overcome by the bridging courses as described. One engineering professor added that it was necessary to modify the two-year common course by taking out the prospective long course students in the second year for additional theory work 'which they need and which would be too demanding for the others'.

As indicated above, the response of employers to the *Gesamthochschule* graduates and their qualifications is as yet unknown. But this response will be of great importance, because until the opening of the *Gesamthochschulen,* the qualified engineer in Germany would be either Dipl. Ing. or Ing. Grad. licensed by the state educational system and not by a plethora of professional associations as in Britain. This was straightforward and easily understood; just how employers will react to Dipl. Ing. (long course) and Dipl. Ing. (short course) as well as the traditional qualifications remains to be seen.

There is the possibility that the status striving of the students themselves may upset the egalitarian hopes of the educational reformers. One obvious manifestation of this is that the overwhelming majority of the students say, at the beginning of the course, that they hope to complete the long course, whereas the Ministry of Education expects roughly a fifty-fifty split between the long and short courses. There may easily be other manifestations of this status-striving which are not easy to anticipate. For instance two completely parallel common courses are being run, one by a former university teacher and one by a former *Fachhochschule* teacher, if left a free choice, will the majority of students choose the former university teacher's class in preference to that of the former *Fachhochschule* teacher? This happened in the ratio of 100 : 20 in one of the economics departments running parallel courses.

The integration of the two different sets of staff seems to be the major problem since there are distinct differences between them:

1. All are university graduates but few of the *Fachhochschule* teachers have Ph.D.s, whereas the university teachers have, or will eventually have Ph.D.s and the *Habilitation* (a post Ph.D.. thesis qualification, and pre-requisite for a professorship).
2. The university teachers research and publish, the *Fachhochschule* teachers on the whole do not.

3. The university teachers have assistants (because they do research) whereas the *Fachhochschule* teachers do not.
4. The university teachers hold the higher positions, have a higher status, and generally 'run the show'. That is, they have a determining voice in respect of course content and assessment. They, and only they, are eligible for the higher administrative posts of Dean and *Rektor*.

    It should be added that the law which set up the *Gesamthochschulen* in *Nordrhein-Westfalen* also decreed that all members of the teaching body should have university status. This part of the law was subsequently declared unconstitutional by the German Constitutional Court, but it was said that the *Fachhochschule* teachers were hoping for a political move which would reverse the decision of the Constitutional Court, and give them university status after all.
5. There is a wide difference in the class-contact hours: for the former university teachers it is roughly eight hours per week, but for the former *Fachhochschule* teacher it is sixteen. There did not seem much hope of reducing this inequality, because to bring down the class-contact hours of the former *Fachhochschule* teachers would mean doubling their numbers and this could not be afforded.
6. The former *Fachhochschule* teachers are experienced in teaching engineers for the Ing. Grad. qualification but not at university level for the Dipl. Ing.; the reverse holds for the former university teachers.
7. Whereas the former *Fachhochschule* teachers are over fifty and have lived in the area of the new *Gesamthochschulen* for a long time, the staff recruited from universities are relatively young, even the full professors, and are probably newcomers to the area. Thus the organizational differences are reinforced by a locals-and-cosmopolitan split.

A final point to be considered is student motivation in choosing where to study. It is understandable that students without the *Abitur* qualification will choose a *Gesamthochschule* in preference to a *Fachhochschule* because it opens up the possibility of doing degree work. But why do students with the *Abitur* choose a *Gesamthochschule* rather than a university?

Two explanations were put forward. First if there is a *Gesamthochschule* but no university in their home town, many students will choose the *Gesamthochschule* and stay at home, because it is cheaper and it is traditional in Germany to remain at home during the time of higher education. The second explanation was that the student has little choice, being directed to a *Gesamthochschule* by the ZVS (*Zentralstelle für die Vergabe von Studienplätzen*) (see page 21). In this case although the students are directed in the first place, they have the right to transfer and many do. Such a reaction upsets the intentions of the educational planners who anticipate both a fifty-fifty intake of students with or without *Abitur* and a fifty-fifty split between the long and short courses after the first two years.

It is interesting to note from our survey that of those who did not have a university education, fewer than half said that they would have liked the opportunity to have gone to a university. Table 2.1 shows the most common reasons for not going.

TABLE 2.1

Lack of money	40 per cent
No Abitur	15 per cent
Family circumstances	14 per cent
1939–45 War	13 per cent

We would conclude by saying that none of the problems indicated should be regarded as insuperable. The *Gesamthochschule* movement is still in its infancy, and has not been helped by financial stringencies. Passing judgement at this stage would be like passing judgement on English comprehensive schools in 1955.

Chapter 3

Recruitment and Attitudes to Education

The recruitment of engineers is a vital aspect of the engineering profession and indeed the national economy. A study of where they come from, when they choose their career, and why, should provide valuable pointers to the factors influencing the distribution of manpower within the economy. How should a country recruit not only enough engineers, but also attract the best quality? Are there social barriers such as class, wealth, family background, which impede the optimum recruitment process and how does schooling affect subsequent career? These are some of the fundamental questions which come at the beginning of any occupational analysis. Also important are the attitudes of mature engineers to the education and training that they have received and its relation to their later accomplishments. Has the country's educational system served them well and can it be improved or modified to meet the demands of today? This seems to be a subject, like medicine, on which all laymen are prepared to speak at length, but we feel that it is much safer and more objective to find out by means of a survey the opinions of engineers based on their own experience. The results of our survey indicate some of the answers to such questions regarding engineers in the German Federal Republic which will be compared with a similar study of British engineers.[1]

DEMOGRAPHIC AND SOCIAL FACTORS

Compared with Britain a higher proportion of West Germans live in small towns and villages and it is not therefore surprising that only about one third of the German mechanical engineers in our survey were born in large cities, more often these being Dipl. Ing. than Ing. Grad. Most (21 per cent) came from *Nordrhein-Westfalen*, the biggest industrial area, followed by *Nieder Sachsen, Baden-Württemberg,* and Bavaria (all about 10 per cent). One quarter were born outside West Germany.

Perhaps as a result of the numbers killed in the last war, and the after effects of the war, the German engineers tended to be slightly younger than British engineers. Fewer than one in ten were single and only one in 500 was a woman, and the proportions of Protestants and Catholics were similar to the proportions for the whole population.[2]

Their backgrounds were predominantly middle-class, particularly the university graduates, and a higher proportion than in Britian on both husband's and wife's side came from professional executive families with a correspondingly lower proportion from white-collar and manual worker's families. This reflects the generally higher status of the engineering profession in Germany and suggests that sons of professional families are more attracted by engineering as a career than in Britain.

When asked to say in which social class they now considered themselves, 93 per cent of the Germans claimed at least middle middle-class status although in fact only 35 per cent were born to it. University graduates, those with good degrees, and the successful, were all more ready to describe themselves as upper-class. This tends to support our hypothesis that Germans take an achievement oriented rather than an ascriptive view of social status. Their line of reasoning would appear to be 'If you have a good degree and a high income you must be upper class'.

Engineers in both countries showed the modern European tendency to strong upward mobility in educational terms, but the type of secondary school attended had a major influence on going to university or not. Nearly all had higher educational qualifications than their parents. About the same proportion (4 per cent) in Britain and in Germany had fathers who were engineers, but in Britain a higher proportion of engineers had fathers who had been to university, not necessarily to study engineering. On the other hand, as the German engineers at all levels were more predominantly grammar school products it could perhaps be said that as a group they were more broadly educated at school. About the same proportion of engineers in both countries had been to university, but of the remainder who had not the Germans could be said to be better educated and trained.

ENGINEERS AT SCHOOL

In both countries psychological studies[3] of teenage grammar school pupils have shown that future engineers had certain distinguishable group characteristics.

Bartenwerfer and Giesen described a very interesting survey of seventeen-year-old *Gymnasium* students in Germany comprising 284 girls and 434 boys. They carried out a detailed study of the attitudes, interests, and school performance of different groups intending to study seven main disciplines at university, including engineering. Among the boys this was the second most popular choice after mathematics and science.

Their performance was measured by various tests, spatial perception, mathematical, figure, and visual reasoning, and by school performance. The results can be summarized as follows:

1. The would-be engineers were good at, and interested in everything relevant to the later study of engineering. Their showing on mathematical tests was good, and often better, than that of the intending mathematics and natural science students.
2. The 'engineers' were poor on verbal fluency, uninterested in languages, history, and literature, and their performance in German as a school subject was poor. This lowered their achievement over-all in the combined school subjects' mark.
3. They had narrow interests but high ability in mathematics, science, and technology, and little else, but their general knowledge was less restricted than their interests.
4. They were introverted, and uninterested in social welfare questions. They had an average rating on dominance and achievement drives. They viewed school positively and favourably, and were keen on material prestige considerations, and saw themselves 'ready for action'.

Such attributes may be general characteristics of engineers in many other countries. Certainly evidence from Britain and the USA confirms the existence of similar tendencies. Two points, however, should be underlined. The high level of attainment in maths and natural science subjects probably contrasts with Britain, and the very positive attitude to school and what it has to offer is in line with their strong careerist stance.

In Britain Liam Hudson[4] for many years made psychological tests of public schoolboys ranging from 14 to 18 years old,
and found that, broadly speaking, the future engineers and
scientists among them were less flexible intellectually than
those who would follow arts subjects, and were more restricted emotionally. Whereas the arts group were strong on
verbal tests and relatively poor on numerical and diagrammatic tests, the engineers and scientists scored well on
diagrammatic and numerical tests, but were weak verbally. As
the title of his book suggests, they had 'contrary imaginations'
and perhaps personalities too. Certainly Hudson showed that
there were mainly two types of thinker, the convergers (engineers and scientists) and the divergers (broadly the arts students). There was also a small third class of people who could
go in either direction.

In the USA Gerstl and Perrucci[5] found similar tendencies and
differences in attitude. They quoted studies of college students
in the USA which showed engineering students to be low on
interest in people, high on technical self-expression, and high
on making money. They also suggested that 'The combination
of an early career choice and a career decision based upon
interest in the substance of engineering activities or the financial and career possibilities of engineering could lead to an
indifference to all activities that might be considered as distrac-
tions from their main interests.' This point may also help to
explain why many engineering students have little interest in
the social sciences and humanities; and why they devote little
of their time to social and political activities.

In both Britain and Germany engineers, particularly the
most able and those who were successful later in life, had
shown an early technical interest and ability. Most decided to
become engineers when they were between the ages of fifteen
and eighteen, but a higher than average proportion of those
with good degrees and those who became successful decided at
less than fourteen years old. Early motivation therefore seems
to have been a key factor. The influence of parents, friends, and
parental occupation was also noticeable but was stronger in
Germany than in Britain. Dipl. Ings. tended to have been more
influenced by their interest in natural sciences and by having
fathers who were engineers, whereas Ing. Grads. had been

attracted by the type of work done by engineers and by the better career prospects offered. An unusually high proportion (85 per cent) were completely satisfied with their choice of profession. Sixty per cent of the whole sample had spent some time at *Gymnasium* about half of these obtaining *Abitur*, but there were relatively fewer of these in private industry. All but 10 per cent of the Dipl. Ings. and about half the Ing. Grads. had *Abitur*. Only 14 per cent had been to comprehensive schools.

The average school leaving age for Dipl. Ings. was 19.1 years and that for Ing. Grads. was 17.8 years. The Dipl. Ings. had spent about a year longer learning general subjects and languages at school than had the Ing. Grads. They also had higher grades in general in natural sciences and in mathematics than the Ing. Grads. Managers when at school had average academic ability whereas the Dipl. Ings. and those with good degrees had been above average at school.

Because the *Gymnasium* course is longer than that at the English grammar school, the German school-leavers entered university about one year later than in Britain, they were also, on average, about 1.3 years older than those entering *Fachhochschule* to take the Ing. Grad. course and had spent longer studying general subjects. The *Abitur* system also gave a broader education in at least five subjects compared with the narrower two-or three-subject 'A' level system and all German *Abiturients* had studied mathematics, German, and a foreign language to the age of nineteen. They could therefore be said to have received a broader education than their British colleagues.

SCHOOL TO UNIVERSITY

There is considerable evidence to show that engineering in Germany attracts a higher proportion of top-quality students than in England.

Using English university entrance statistics, one of the present authors[6] has shown that the subjects which attracted the very best students, judged from their 'A'-level scores, were medicine and law, followed by the humanities — essentially classics, mathematics, and to a lesser extent physics. All engineering subjects, with the exception of chemical engineering, attracted few top-class students. Practical and vocational subjects generally, with the exception of medicine and law, were also less favoured.

Although corresponding quantitative data were lacking in Germany, it seemed that there was reason to believe that the *Abitur* attainments of German engineering students were better relatively than the 'A' levels of their English counterparts. Engineering subjects also attracted a higher proportion of the total student enrolment in Germany than in Britain (26 per cent compared with 21 per cent). It should be noted however that, because only about half the engineering students in Germany survive their course at university there was a natural selection process towards better quality people during the six years at university. Although therefore because it is a selective system, competition to enter university in Britain is tougher, it may nevertheless be easier for the less gifted academically to survive a three-year course once they have got in.

Also, comparing engineering with other disciplines such as medicine and law which, in both countries, also attract good students, it seemed that in Germany the quality of the engineering intake was relatively high.[7] In Britain the tradition has been that the best mathematics /science pupils at school read science or mathematics at university, whereas in Germany a higher proportion of the gifted students choose engineering as a career. The result is that whereas in Germany for every scientist there are four engineers, in Britain the ratio is one to two. In this respect Britain is exceptional compared with the other western countries in having such a high proportion of scientists and it certainly suggests that British engineers do not form the academic élite as they do in France and to a lesser extent in Germany and Scandinavia.

There are corresponding differences in the university populations of both countries between engineers and science students. Whereas the proportions of engineers to the total student population are not dissimilar, the proportion of scientists in British universities (41 per cent of the total) is exceptionally high compared with Germany (25 per cent). There was also a much higher proportion of British students studying the humanities (24 per cent) compared with Germany (8 per cent), whereas conversely in Germany a much larger group studied law and economics (41 per cent) compared with Britain (15 per cent). This is because the choice of discipline is much more strongly related to vocation in Germany. For instance the main jobs for history graduates would be in teaching and very little else; similarly graduates entering the German Civil Service require a law degree, rather than the wide range of possibilities available in Britain.

Nearly 90 per cent of German engineers had graduated since 1950 half of them between 1960 and 1970.

Academically the quality among the older German engineers who had graduated between 1940 and 1950 tended to be much higher than among the younger generation. Perhaps the post war students of 1946 onwards were exceptionally good or at least particularly well motivated. Certainly a high proportion of that group are now doctors of engineering.

In their school subjects the Dipl. Ings. had obtained higher average grades than the Ing. Grad. and in their engineering degrees a much higher proportion of Dipl. Ings. than Ing. Grads. had obtained top class grades. It seems therefore that a higher proportion of Dipl. Ings. were academically gifted and well motivated, the latter perhaps assisted by family background. An apprenticeship had been served by 80 per cent of the Ing. Grads. and 24 per cent of the Dipl. Ings. and most of these felt that this time had been well spent. A lower than average proportion of ex-apprentices was found among those with good degrees, the successful and the managers. Most of the Ing. Grads. would have done their apprenticeship between leaving school and entering *Ingenieurschule* under the old system. As they had spent an average of 5.7 and 3.4 years on university and college courses respectively the average total periods for those who had done apprenticeships were 8.3 and 6.3 years for the Dipl. Ings. and Ing. Grads. respectively. The relatively high proportion of Dipl. Ings. who had served an apprenticeship is noteworthy and also the fact that 6 per cent of the Dipl. Ings. had first obtained the Ing. Grad. qualification. There was thus an appreciable proportion of university graduates who had 'come up the hard way' and who could therefore call themselves good practical men.

ATTITUDES TO HIGHER EDUCATION AND TRAINING

Not only is the type of education and training received by engineers likely to influence their subsequent careers but their attitudes towards their early experiences may affect their development. A comparison of the attitude of German and British engineers is therefore of interest. It should however be pointed out that there was a time interval between the two surveys now compared of about twelve years. Intervening changes in education and social conditions could therefore influence the validity of comparisons. Nevertheless in view of the conservative nature of engineers, the relatively slow changes in curricula, and experience of other surveys in USA it seems that there is much common ground.

The views of engineers in the two countries on their education and training were remarkably similar although the systems differ. German university engineering education is more specialized and involves an average stay of six years which includes six months industrial training, whereas in Britain by contrast, only three years are spent at university plus two years training in industry for those who seek the Chartered Engineer status awarded by the Council of Engineering Institutions. In Germany the Ing. Grad. used to do an apprenticeship or two years industrial training on leaving school, then three years study at engineering school, but he now spends three years at *Fachhochschule* after a two year full-time preparatory course, including some practical work at a junior technical college. Academically the Ing. Grads. could be compared, roughly, with British engineers having a Higher National Diploma plus industrial training.

Most German engineers were satisfied with their education and training, but there was an appreciable feeling among the Ing. Grads. that their courses had not been practical enough and were too diluted and superficial. Most engineers would like to have studied additional technical and non-technical related subjects such as business-economics, management and operational research, and a foreign language, all rather more favoured by Ing. Grads. whereas the Dipl. Ings. preferred political economy. In both countries the most popular additional technical subject was electrical engineering. Their views on the objectives of an engineering course were similar but the Dipl. Ings. ranked critical thinking and the study of scientific method a little higher than did the Ing. Grads. who in contrast rated business studies higher.

Comparing attitudes towards university engineering education in both Germany and Britain most of the non-graduates would like to have gone to university, but for those who had to study elsewhere their main obstacle to university education had been lack of education and money which were related to family background. The non-graduates gave higher status and prestige as the main reason for preferring a university education whereas the graduates believed that their technical education at university had been better. They agreed that the greatest need for a university education was among those working in research and development.

In both countries about half the non-graduates said that they never had wanted to go to university and that missing it had made no difference to their careers. However there was a slightly higher proportion wishing to have gone to university among British engineers. About one sixth in each country thought that the lack of a university education had hampered their career and therefore had strong regrets. In fact about 44 per cent of Ing. Grads. in general had no regrets. As the proportion having no regrets was greater in Germany than in Britain, it seems that the importance of a university education is not quite so great, perhaps because Germany is more of a meritocracy and less class-conscious than Britain. Of those who claimed that it had made no difference to their careers the largest group were managers, but of those engaged on purely technical work about one fifth said that they could progress no further without a degree.

There was also evidence that German engineers in their student days had a stronger technical as opposed to management orientation than the British and yet paradoxically in later life a higher proportion of them became top managers. The route to the top in Germany therefore seems to be for those of proven ability in a technical job, rather than for generalists, as it is in Britain. It is assumed and proven by experience in Germany that outstanding engineers are also good managers. Career ladders in strictly engineering functions also seem to be longer in Germany than in Britain. In contrast the British see that to secure a well paid job it is necessary to become administrative and, that to reach the top, it may even be necessary to be trained in the subject of management which the Americans invented and which the British, of all the Europeans, believe in most devoutly. Another example of strong technical orientation is the remarkably high proportion of German engineers who had served an apprenticeship and were qualified craftsmen.

About a quarter of the German university students thought their course had been too long and detailed whereas one third of the Ing. Grads. thought theirs had been too short and superficial. There was more dissatisfaction with industrial training in Britain but although most of the Germans were satisfied, some thought that they had spent too long on routine production, particularly the university graduates and those generally who became managers.

Despite the much greater length of the university courses in Germany about 10 per cent of Dipl. Ings. had stayed on to obtain Doctor Ing. degrees which is, incidentally, similar to the proportion of engineering Ph.D.'s in the USA but about three times the proportion in Britain. It seems therefore that Britain has fewer specialists than either Germany or the USA.

It also seemed that Dipl. Ings. had a much higher output of publications, lectures, and patents than their British counterparts (see Chapter 10). Whether this is because of different personal qualities, education, or incentives is not clear but the difference was remarkable.

THE IDEAL ENGINEERING COURSE

Views on what represented the ideal curriculum were remarkably similar between graduates and non-graduates in both countries. The British and German samples had been asked several questions about their usage and rating of many subjects studied at university and college. Among these was the question 'What percentage of time, if any, would you recommend should be devoted to each of these areas in an engineering course?'

The replies to this question summarized in Table 3.1 show broadly similar views about the proportions of related non-engineering subjects such as humanities, languages, industrial administration, economics, and social sciences. The main difference is that of balance between basic engineering science and design engineering. Germans have a strong tradition for

TABLE 3.1
*Recommended Allocation of Time in an Ideal
Engineering Course*

| | percentage of total time | | | |
	GERMANY		BRITAIN	
Humanities	10		7	
Foreign Languages	9		7	
Technical Report Writing	7		7	
Industrial Administration, Economics Social Science	12	38	10	31
Fundamental Sciences, Mathematics		23		23
Basic Engineering Science		3		27
Design Engineering		24		13
Speciality Engineering (e.g. Textile Engineering)		11		6
		100		100

design whereas the British place more emphasis upon engineering science.

On the general distribution of time between engineering and non-engineering subjects there was once again remarkably close agreement between Germany, Britain, and the USA where the Grinter Committee some years earlier had published a report on engineering education[8] which made similar recommendations. They all agreed for instance that about 30 per cent of the time should be devoted to social sciences and humanities. Languages and report-writing were allocated about two thirds of this the remainder going to the social sciences of which the instrumental subjects such as economics and law were given more time than the behavioural subjects such as sociology, psychology, and politics. In fact the only country that now devotes such a high proportion of time as 30 per cent to related non-engineering subjects is the USA. Germany and Britain devote the least time (about 12 per cent) and France is in the middle (about 17 per cent). The relative importance of the social sciences is discussed in more detail by one of the present authors[9] who concludes that there is no evidence to suggest that emphasis on management and business studies in an engineering course will produce better managers. It is more important to ensure good all round abilities and that top quality entrants are fed into the system. No doubt the German engineer's dominance of management posts at all levels reflects these facts of quality and general aptitude, as well as the German veneration of the technical.

Chapter 4

Employment and Career

Having invested much time, energy, and money in educating the young generation of engineers, it is important to know how these assets are disposed in space and time over the country. Where do they all go, what do they do, and how do they fare? Or to put it more within the framework of our questionnaire when the survey was carried out, where had they been, what had they done, how had they fared, and what did they think about their future career prospects? It is a stock-taking process of a dynamic and ever-changing phenomenon, and it is not easy to sort out from the stills what is happening in the motion picture. However let us try to analyse the situation from the results of our questionnaire survey.

At that time in the German manufacturing industry there were about three Ing. Grad. to every Dipl. Ing., that is about one quarter of the engineers had spent six years at university. There is evidence that this proportion among the young generation is increasing, because more who start on the Ing. Grad. route are tending to transfer to or proceed with the Dipl. Ing. In Britain too the proportion of university-trained engineers has been increasing. In 1962 about one quarter of all engineers, as in present-day Germany, were university graduates; but with increasing opportunities for going to university the proportion of graduate British engineers has today risen to nearer 60 per cent and it appears that there are similar social and educational changes occuring in Germany despite the longer six-year university course, double that of the British graduate.

The possibilities for employment in post-war Germany have been vast because of the steadily expanding economy. Only during the few years of reconstruction immediately after the end of the war could it be said to be difficult, and after that the manufacturing industry expanded rapidly. In such conditions, therefore, it has not been difficult for graduates to find work, and they have experienced a free market for labour. It is therefore very interesting to see in which directions they have gone in this type of industrial climate.

Because nearly all students had completed their industrial training before or during their course they were able to start work as professional engineers upon graduation. A few however did go on at this stage with further training, but two-thirds had done no further in-house training lasting six months or more when they began their first job. Of the third who had received additional training the highest proportions were those who had become managers, those working outside private industry and those with good degrees. As those most seldom receiving further training had below average degrees, it seems that those selected for further training had been expected to have technical or managerial potential.

The decade 1960–9 was the time when half the sample began their career, with the other half divided almost equally before and after this time. It was a period of social change and student unrest, which is perhaps reflected in the high proportion of below-average degrees.

EARLY CAREERS

For their first and two subsequent jobs nearly one third of the engineers were working in *Nordrhein-Westfalen*, with lower proportions in descending order, in the *Länder Baden-Württemberg*, Bavaria, Lower Saxony, and *Hessen*. Relatively more Ing. Grads. than Dipl. Ings. were working in *Nordrhein-Westfalen*, while the reverse was true in *Baden-Württemberg*; this was noticeable for jobs two and three as well and perhaps reflects the different types of industry in these areas.

Of six categories of organization, viz. industry and commerce, state enterprises, consulting, large research institutes, public corporations, and universities and *Fachhochschulen*, by far the largest proportion (80 per cent) of the sample began work in industry and commerce, with the steel industry, machine and vehicle construction firms employing more than half of them. Very similar proportions were found for jobs two and three with never less than three-quarters working in industry and commerce.

The analysis of which particular firms the newly-qualified engineers joined was unsatisfactory, because the coding of the survey concentrated on certain big firms which seemed to account for only 25 per cent of the respondents. There was certainly no large concentration in any one firm, about one sixth of them were spread evenly among eight large organizations, of which the Post Office and Siemens took the most. These two were less popular in subsequent jobs but there was still no marked concentration. The answers regarding size of the firm were more satisfactory, one quarter of the newly qualified engineers joined work units of more than 5,000 employees, but a higher proportion than the sample average of Dipl. Ings. and those with good degrees went into the smaller units of less than 100 employees. There was a tendency for people to move towards the smaller firms in their second and third jobs.

There seemed to be a definite polarization between the type of job and the qualifications of the new employee. Although the largest single group were employed in design, it was much more the province of the Ing. Grad. and those with below average degrees. In contrast those with good degrees started more often in research and development. This polarization tended to persist in subsequent jobs although not quite so strongly, and was the most marked difference found between the two types of engineer. For instance twice the proportion of Dipl. Ings. first went into a research and development job and, in type of work, three times the proportion of Dipl. Ings. were involved in work concerned with research and development compared with Ing. Grads.; on the other hand twice as many Ing. Grads. were employed on design.

MOBILITY

It is remarkable that over a third of all the engineers had not moved from their first firm and that just under a third had made only one change. This means that two out of three of them had not felt that wider experience or possibly better chances of promotion in a variety of jobs would enhance their careers.

It was to be expected that a high proportion of those still in their first job would be in the younger age group, but it is surprising that as many as 39 per cent of the 35–49 age group had not moved at all. Of the more mobile engineers those in management services or non-profit-making organizations were more likely to have had three or four jobs and those in universities and *Fachhochschulen* were more likely to have had six jobs, but these groups represented only a very small proportion of the total. Of the remainder the average stay with a first employer was about three years. Those who were between forty and forty-five at the time of the survey had tended to stay longer in their first job, perhaps because at the time when they began there were fewer opportunities for moving.

The most commonly quoted reasons for moving were better promotion prospects, immediate job enhancement, and better chances of the job content improving. Those moving frequently tended to have a higher average salary, but among the successful (i.e. the top quartile for age group) there was no correlation with number of job changes, so for the top jobs it seems just as worth while to stay with the same employer. It is interesting to note that they are much less mobile than their British colleagues.

CAREER PATTERNS

From all the detailed information received about jobs, it was evident that the changes in distribution and deployment of the engineers with time was not very great, and a definite pattern therefore emerged. Industry was the largest employer, with a relatively higher proportion of Ing. Grads. than Dipl. Ings., and of the successful. The largest group, approximately half, worked in the steel industry, machinery, and vehicle construction. In the non-profit-making organizations there were more than double those with good degrees compared with below average degrees; the electrical engineering industry also attracted those with good degrees, while the steel, machinery, and vehicle construction industries had a high proportion of those with below average degrees. The proportion of engineers working in state enterprises hardly changed and always combined relatively larger proportions of the unsuccessful and those with below average degrees.

Between the first and the present job there had been a movement from private industry into consulting and testing services and into the universities and *Fachhochschulen*, particularly those with a Dipl. Ing. As regards type of work there was always a higher proportion of engineers working in design than in any other function, although it had decreased since the first job. Between the first and present jobs also there was a small decrease in the proportion working in research and development and an increase of those in testing and inspection and technical administration, the latter probably being an age-linked seniority trend.

It was noticeable that the German engineers tended to stay not only with the same employer but also in the same field of work. At the time of the survey 84 per cent of those working in design and 74 per cent in research and development did this type of work in their first job. On the other hand, just under half who had started in a testing and inspection field were still in it, the vacancies being filled by people changing from design work or to a lesser extent research and development.

As professional engineers one third had never been called on to do any manual work and only one fifth had often done any and that was most frequently in their first job. There was general agreement that such collaboration with manual workers was useful, but one-tenth, particularly among the Dipl. Ings., were against it.

During their careers about half the German engineers had been abroad on business trips and about one in ten had worked abroad for a length of time. Linguistically the Dipl. Ings. were better than the Ing. Grads., as they should be from their compulsory study of at least one foreign language up to the age of nineteen; half of the sample claimed to speak a foreign language, more often English, fluently.

PRESENT JOB

Having examined the general patterns and trends that have developed with age and career, now lèt us see what people actually do in their present job, how they spend their time, how much authority they have, and what job satisfaction they derive. This latter is, not surprisingly, connected with how much money they earn, although Germans are rather reluctant to discuss their salaries. Another more philosophical facet is what plans they have for the future, how long do they intend to stay with their present employer, and what they expect their career peak to be. So let us examine the facts of working life and what the attitudes and expectations of engineers are.

When the engineers worked out the percentage of their total work time spent on various activities, it was found that design absorbed the most, followed by testing and inspection, technical administration, research and development, and routine office work; these varied in descending order from 15 to 10 per cent. In a slightly different way the sphere of work where the majority of engineers had full authority to make decisions, the order was routine office work, testing and inspection, technical administration, design, and research and development ranging from 71 to 50 per cent of the whole sample. Relatively more Dipl. Ings. were involved with non-technical aspects then Ing. Grads. who tended to be more technical. The three types of work where engineers seemed least often to have full authority were costing and finance, operational research, and research and development. It appeared that a larger proportion of engineers had authority to spend money (1,000 DM) than to appoint qualified or office staff. Although one-quarter had their own secretary and one-half shared one, most of the sample felt that some of their work, particularly routine office work, could be done by someone with less technical training. Even so they were much better off than the British engineers, of whom only 55 per cent had any secretarial help.

As far as work place was concerned more than half worked in a complete industrial unit rather than an outstation and of these there were a high proportion of the successful and those with good degrees. The average number of engineers and scientists employed at the work place was eighty-three. Those with good degrees tended to avoid large units (100 or more) and were most frequently found in units ranging from 10 to 19 and 20 to 49 qualified colleagues. The successful on the other hand were found in large units and least often in smaller ones suggesting that remuneration may correlate with responsibility for people. The Ing. Grads. tended to be found more in the smallest organizations and in all other sizes the Dipl. Ings. were in larger proportions.

As far as close working relations with qualified personnel were concerned, the average was a group of ten. About one-quarter of the engineers had no responsibility for other staff and one-quarter were in charge of up to five qualified colleagues.

'Most of the sample were satisfied with their work and working conditions: 80 per cent had a prescribed forty-hour week, although they actually worked longer hours than this, but 10 per cent, mostly the successful who were presumably their own bosses, had no set hours. Only 55 per cent were covered by a pension scheme, by far the most common type being contributory, and only 10 per cent had a civil service pension.

REMUNERATION

In the course of the survey the engineers were asked what their monthly salaries were, and although four years have elapsed since then, four years of mounting inflation which make the actual figures seem low, it is still useful to look at them in their present-day context.

The salary scale ranged from below 1,999 DM to over 7,000 DM. Over two thirds of the sample lay in the 2,500–4,500 DM bracket, that is when converted at the then current rate of exchange, they had a gross monthly salary of between £625 and £1,125, which would give an average of £11,187 per annum. This can be compared directly with a salary survey conducted at practically the same time by the Institution of Mechanical Engineers.[1] Among a sample of 5,700 members of the British institution the median income was £5,510 per annum, half that of the German engineers. There are two factors which make this difference even greater. The first is that income tax rates do not rise as high in Germany as in Britain, hence a higher gross salary results in a higher net income than it would in Britain. The second factor is that the monthly salary should be multiplied by thirteen, not by twelve, to convert it to an annual sum, since salaried German engineers will normally receive thirteen monthly payments per year, the thirteenth being in the nature of a 'Christmas' bonus; this makes the German average, as of March 1976, £12,199 per annum. On the minus side the cost of living is somewhat higher in Germany than in Britain; for 1976 the UN index put it at 36.4 per cent higher, but even allowing for this it is clear that the average German engineer earns considerably more in real terms than his British colleague.

At the very top, 2 per cent of the German engineers were earning over 7,000 DM per month or the equivalent of

£22,750 per annum, and at the same time only 0.8 per cent of the British engineers earned over £10,000 per annum.

It was evident that the better qualified engineers earned more than their colleagues with lower qualifications. Nearly half of those with a Dipl. Ing. were earning the equivalent of at least £1,000 per month, but only a quarter of those with Ing. Grad. were in this group. There was an even higher proportion (62 per cent) of those with Dr. Ing. in this salary bracket, but as the actual number of Dr. Ing. in the sample was small, too much should not be claimed from this figure. This fundamental factor also emerged very clearly in the earlier survey of British engineers[2] and was again evident in the comment on the Institution of Mechanical Engineers survey by S. P. Hutton.[3]

Another interesting point noted in relation to the basic qualifications of engineers was the degree of overlap in salaries, that is, no matter how high one goes in the Dipl. Ing. salary scale there were always some with the Ing. Grad. qualification at the same level.

By calculating salaries for all engineers working for one particular type of employer, e.g. industry, university, etc. it was possible to see the relative levels of remuneration for each type of employment. Those with the top salaries were working in industry or its associate consultancy, the middle group worked with public utilities or in administration, and the lowest were engaged in academic work. This is completely the reverse of the situation in Britain where engineers working in industry are the worst paid. It is also significant that those engaged in academic work are the lowest paid among German engineers in the light of the previous finding that the better qualified earn a higher salary than the lesser qualified. It certainly does not derive from the relative absence of those who have a Dipl. Ing., quite the contrary, because university graduates are strongly represented in academic work, but from the fact that other types of employer, notably industry, pay more. Nevertheless it should be remembered that engineering professors and senior lecturers in Germany are able to augment their basic salaries by carrying out contract research and consulting work for industry. The higher salaries earned in industry are consistent with the claim, which will be discussed in more detail later, that industry has a higher standing in German society.

THE FUTURE

When asked about their plans and expectations for their future careers, nearly half the sample replied that they had no further plans, this being particularly noticeable among those who were teaching. Ten per cent were aiming at becoming head of a department, and another 10 per cent were planning on getting promotion, this was strongest amongst those employed in state enterprises. Only 2 per cent were planning to change jobs.

Another question was what position they expected to be in at the peaks of their careers, and again the largest proportion (29 per cent) were negative, they did not know; 22 per cent thought that they would be head of a department and 14 per cent felt that they had already reached the peaks of their careers. On the positive side the majority thought that they would reach this career peak while with their present employer. This was confirmed by parallel questions of how long they expected to stay in their present jobs and with their present employer, when 41 per cent expected to remain until they retired. Bearing in mind their age distribution (42 per cent were 40 and over) the engineers seemed to be remarkably settled in their jobs and even more so with their employer.

Considering the relative immobility and the dearth of plans for the future, it is not surprising that the majority of the German engineers expressed a high degree of satisfaction with their jobs, although it should be noted that there were 10 per cent who were not particularly satisfied by any aspect of their work. They particularly liked the self-reliance, independence, and ability to organize their own work, and 10 per cent mentioned the satisfaction of design and creativity; the most unsatisfying aspects were administrative and routine work.

To all these questions the replies of the Dipl. Ings. and Ing. Grads. were remarkably similar; one difference worth noting was that 4 per cent more Dipl. Ing. than Ing. Grad. did have plans for the future, and slightly more than the sample average hoped to become Directors.

Chapter 5

Successful Engineers

One effect of the rise of sociology is that we have all become accustomed to looking for structures and causal patterns. In this climate the quip 'success is all a matter of luck, ask any failure' loses some of its sting, in that we would at least expect 'luck' to have its own consistencies.

So a principal reason for examining those members of an occupational group who are especially successful in their working life is an interest in the causal. A desire to know why these people, and not some others, succeeded occupationally. But this is not the only reason for an interest in the successful minority: it is also a question of characterization. What, in short, are the successful like; what are the manifestations of success, how are the successful distinguished, what is different about their contribution in the occupational sphere in which they are deemed to have succeeded? To this idea of characterization may be added that of exemplification. As successes the successful must really typify their occupation, showing its nature and dynamics in pristine form. What is more the whole question of who succeeds under what conditions and why, should also tell us something about the society in which it is happening.

This last consideration, the dynamics of occupational success as a reflection of society's values, has considerable potential interest in the case of Germany. This is partly because Germany does have a popular reputation for being 'good at making things', for exalting in the technical, and partly because Germans themselves tend to express the view that their society is a particularly meritocratic one in which *Leistung* (achievement, performance) is what counts. If this is so, it should be discernible in the analysis of a minor élite group. One also hears more often in Germany than in Britain the idea that the country has become 'a classless society', and this proposition is interesting as ideology even if it does not square with the facts.

These remarks presuppose that there is some criterion for

defining success in the present context, and some mechanism for isolating it from the average performance. In fact it is difficult to develop a good mechanism in the case of an occupational group such as engineers, who are so heterogeneously employed. Consider, by way of converse illustration, how much easier it would be to select out the successful from a larger sample of military officers. One would be justified in deeming all holders of field rank as successful by the standards of their calling, or, if one wished to have a more age-diverse sub-sample of successful officers, it would be possible to designate as successful all those who attained particular ranks at a younger than average age — thirty-year-old majors and so on. In short, the army offers the advantage in this connection of being a monopoly employer with a uniform set of ranks. With engineers, however, whether British or German, there is a variety of employing organizations, in both the public and private sectors, and a small minority of qualified engineers are self-employed as well. Furthermore, the industrial companies which employ the majority of engineers are of widely varying size, especially in Germany where the smaller companies more often employ qualified engineers than is the case in Britain, and they have different structures and hierarchies. So one is obliged to find a definition of success which cuts across all this.

Faced with such complications, we have adopted a salary-based criterion of success for our sample and linked it with age. That is to say we have deemed occupationally successful those engineers in our sample whose salary puts them in the top quartile of the actual salary distribution for their age group, age groups being in five year spans. Thus, for example, if a 37 year old engineer earns more than three-quarters of all the other 35–9-year-old engineers in the sample he is judged successful for the purposes of the study.

This method is not perfect, and three objections come to mind. The criterion of salary is by definition secondary, so that one hopes to be singling out not performance itself but a realistic reward for that performance. Second, the computer operation which cross tabulated success, so defined, with other characteristics, did not for the most part distinguish between those engineers holding the Dipl. Ing. qualification and those holding the Ing. Grad. (graduate and non-graduate qualifications) but the Dipl. Ings. earn 16 per cent more on average anyway. Thus the criterion over-represents the Dipl. Ings., or, to put it another way, Ing. Grads. who come to be classed as successful according to this criterion have performed that much more impressively. The third difficulty is that salaries for engineers are substantially and consistently higher in industry than they are in the public sector, although the gap is thought to have been closing in the late 1970s. So again the success criterion used in this study over-represents those engineers working in industry, who are in any case the majority. This argument is two edged of course. It may be counter-argued that since the disparity in average remuneration between the public and private sectors in Germany is well known, engineers who have more drive and who are 'success-prone' will tend to choose to work in industry anyway. Indeed the comparison between engineers working in industry and those working in the public sector in Germany does give some support to this counter-view (see Chapter 8).

If the salary criterion of success has such drawbacks as these, it also has some advantages. As suggested, it obviates the difficulty of comparing various employment grades in organizations of different size and sector. The criterion is again fairly simple to apply, and enables one to draw a clear line. It has the merit too of a kind of objectivity. The salary criterion, is independent of the views of those to whom it is being applied — the engineers themselves. It is similarly independent of the views and the judgements of the colleagues and employers of those engineers in the sample in a way that, say, relative success ratings of employed engineers by their bosses would not be. The age linked salary criterion has the further twin advantages that it largely eliminates the effects of age, while offering an all-age sub-sample of successful engineers, who are not the oldest, not the highest paid absolutely, but the best paid for their age.

Designating only those in the top salary quartile as success-ful, as opposed to dividing the sample simply into upper and lower halves according to age linked salary, was a deliberate decision. The purpose was to use a criterion that would really differentiate those designated as successful from the rest of the sample, and when success, so defined, is cross-tabulated with other features it becomes clear that this aim has been achieved. It seems that the successful are clearly differentiated from the rest of the sample on many characteristics. As a further check on the analysis of the sample data we also designated those in the bottom salary quartile for their age group as unsuccessful, and then cross-tabulated this status with the response to some of the other items in the questionnaire. On the whole, this exercise has produced consistent results, even if they are not consistently reported throughout the present book. The usual pattern, that is, is for the successful to be above the sample average on a range of variables, and for the unsuccessful to be below the sample average.

To complete this part of the picture of the way the data for the whole sample has been analysed, we also set up sub-categories for engineers in the sample with above average and below average degree or engineering school qualification grades. Both Dipl. Ing. and Ing. Grad. passes are graded on a 1–4 basis, grade 1 being the highest in both cases. So we have classified grade 1 passes (with and without distinction) for both qualifications as above average, and grades 3 and 4 as below average. Engineers with grade 2 passes, the majority in the case of the Dipl. Ings., have been omitted from this operation, again so that we really have sub-categories which can be contrasted with each other and with the sample average. The present chapter is about the characteristics of the successful, not those with good degrees, but at various points we give data for the sub-samples with above average and below average degrees, alongside figures for the successful and unsuccessful, for added interest and comparison. A general consideration is that there is in fact a good deal of overlap between the two sub-groups, those who are successful, and those who have good degrees, not just in the sense that each sub-group contains a higher proportion of the other than it would do if the two variables were not connected, but also in the sense that the successful and those with good degrees share a number of other characteristics.

To begin the characterization of the successful engineers with some general background features it should be said that success does not appear to be geographically determined. The proportions of successful engineers coming from the town and the country, and from big and small towns, is the same as for the sample as a whole. Likewise similar proportions of the successful and the whole sample were born outside the present borders of the German Federal Republic (about a quarter of the whole sample) largely in the areas that are now in Poland and Czechoslovakia, or which form the German Democratic Republic (East Germany). It is a little surprising that this group is not over-represented among the successful, or only very slightly. German folk wisdom certainly has it that these 'Germans from the East' contain a higher than average proportion of (potentially) successful people, because of the survival capacity and/or positive initiative needed to get to the West. The only geographic distinction exhibited by the successful is that their ranks contain a slightly larger proportion than the whole sample of engineers born in the state of *Nordrhein-Westfalen* (the most populous of the *Länder*) and a slightly smaller proportion born in Bavaria.

The relative proportions of Catholics and Protestants among the successful are much the same as for the whole sample, in which Protestants considerably outnumber Catholics. The only difference in the religious affiliation of the successful engineers is that their numbers contain a higher proportion claiming 'no religion' or 'other religion' in the survey, though unfortunately the way this questionnaire item was designed does not enable one to distinguish between Jews, non-conformists, and agnostics.

If no place, and no major religion, is favoured as the cradle of engineering success, a particular period is. A disproportionate number of the successful obtained their engineering qualifications in the 1940–50 period. The consideration here, especially for the second part of that period, is that 'times were hard' and unusually high levels of endurance and purposefulness were required to engage successfully in higher education and training — qualities which had a further determining effect on the subsequent career of those involved.

As one would expect the majority of the whole sample are married, and the proportion of successful engineers who are married is a little higher than for the whole sample. It is, however, a little difficult to interpret this small finding, since its significance would depend on the point in the evolution of their occupational success at which the engineers got married. If the differential marriage rates are difficult to interpret, however, the data on parental background are very clear and do not support the popular German claim to have achieved 'a classless society'.

It becomes very clear that the children of middle class parents have an advantage in life if the whole sample is divided into three groups — approximately working class, lower middle class, and middle class — omitting those respondents who did not answer the question on parental occupation and education, or did not do so precisely enough (see Table 5.1).

TABLE 5.1

Father's Occupation	Successful Engineers per cent	Whole Sample per cent	Unsuccessful Engineers per cent
Professional and executive	47	35	28
Minor officials, technicians and clerks	22	21	29
Manual workers, manual supervisors and farmers	30	33	37

A similar pattern emerges in the occupations of the fathers-in-law of the engineers in the sample:

TABLE 5.2

Father-in-law's occupation	Successful Engineers per cent	Whole Sample per cent	Unsuccessful Engineers per cent
Professional and executive	40	32	28
Minor officials, technicians and clerks	26	27	22
Manual workers, manual supervisors, farmers	28	36	46

A positive association also exists between the parent's educational level and his occupational success. If the engineer's father had *mittlere Reife* (German equivalent 'O' level GCE, see Chapter 2), or anything higher, the engineer's chance of occupational success is increased (see Table 5.3).

TABLE 5.3

Father's educational Level[1]	Successful Engineers per cent	Whole Sample per cent	Unsuccessful Engineers per cent
Secondary modern school leaving certificate	46	55	69
'O' level GCE	24	21	18
'A' level GCE	17	12	7
University degree	5	4	1

The same pattern emerges for the relationship between the educational level of the mothers and the occupational success of their offspring, though the educational attainments of the mothers were more modest overall, with only three of the one thousand or so mothers having a university degree (see Table 5.4).

TABLE 5.4

Mother's educational Level	Successful Engineers per cent	Whole Sample per cent	Unsuccessful Engineers per cent
Secondary modern school leaving certificate	58	66	71
'O' level GCE	28	23	20
'A' level GCE	10	7	4

This relatively favoured background enjoyed by the successful is by no means their only distinguishing characteristic. They were also marked by a stronger commitment to engineering, stronger involvement in their earlier education, and a more favourable *post factum* view of it. The successful engineers, on average, decided in favour of a career in engineering at an earlier age than the sample as a whole. Among those who decided at the age of 14, in particular, the successful were over represented. A higher proportion of the successful had obtained the *Abitur* (German equivalent of 'A' level GCE, see Chapter 2), the successful have higher school marks in general, as well as higher grades in their degrees or engineering school qualifications as already mentioned. They were more likely to have taken additional technical subjects at college (additional, that is, to mechanical engineering), and were more likely to have studied additional non-technical subjects as well.

The successful engineers also tended to exhibit more favourable attitudes to the training and education they had received at the outset of their careers. They were more likely than the sample as a whole to view the Dipl. Ing. and Ing. Grad. courses as 'good' or 'very good', more likely to express a favourable view of the *Praktikum* (compulsory practical training in industry, see Chapter 2), and to judge the over-all length of their course to be satisfactory. The only exception to this litany of favourable evaluations concerns the apprenticeship. The successful were less likely than the sample as a whole to have done an apprenticeship, and more likely than the sample average to criticize some aspect of the apprenticeship. This is perhaps surprising in view of the general esteem in which the apprenticeship is held in Germany. Double apprenticeships (in different trades) are common, and it is not unusual for the Dipl. Ings. to have done an apprenticeship — over 20 per cent of the Dipl. Ings. in our sample have done so.

The whole sample were asked in the survey if they would again elect to study engineering if they could imagine themselves to be at the stage of choice again: the majority said that they would. The engineers in the sample were similarly asked if they would elect to study mechanical engineering in particular: again the majority said that they would. In both cases somewhat higher proportions of the successful expressed this retrospective loyalty to the subject of their originally youthful choice. It is also the case that the proportion of the successful who enjoyed their degree course so much that they stayed at university for postgraduate work is again higher than for the sample as a whole (see Table 5.5).

TABLE 5.5

	Whole Sample per cent	Successful Engineers per cent	Unsuccessful Engineers per cent	Good Degree per cent	Sub-average Degree per cent
Doctor's degree[2]	2	5	0	12	0

In other words there is a positive association between occupational success and having attained the doctor's degree in engineering, though, as one would expect, a stronger association exists between having a good (grade 1) degree and a doctor's degree.

It was observed earlier that the criterion of success used here tends to over-represent those engineers who work in industry. In fact the proportion of those who are successful now, and had their first post in industry after leaving college, is higher than the corresponding proportion for the whole sample (see Table 5.6).

TABLE 5.6

Type of Employer: First Post	Successful Engineers per cent	Whole Sample per cent	Unsuccessful Engineers per cent
Industry	88	80	68

Thereafter, there is a small scale move away from industry in the form of a 10 per cent drift into the public sector. The positive relationship between employment in industry and success, however, is constant (see Table 5.7).

TABLE 5.7

Type of Employer: Present Post	Successful Engineers per cent	Whole Sample per cent	Unsuccessful Engineers per cent
Industry	82	70	54

The functions or areas of work which are associated with success include Production, R and D, Sales, and Finance; on the other hand Maintenance, Testing, Inspection, and Design are negatively associated with success. This omnibus statement is not quite the same as saying that above average proportions of the successful work in the first set of functions — Production, and so on — while below average proportions work in the second — Maintenance etc. This set of associations is derived from statements by the engineers about those areas of engineering work on which they spent varying proportions of their time. In most cases, of course, this will imply working in Production, Design, or whatever. On the other hand the sample will also include some incumbents of general management or administrative posts involving responsibility for several areas of work: here the positive and negative associations still hold, but it will be a case of responsibility say *for* R and D, and time spent on this responsibility, rather than work *in* R and D.

One thing which clearly emerges from the study is that those deemed successful have greater responsibilities. This is not true by definition since the successful are not the most senior but the highest paid in their age group. The whole sample were asked in what areas of work they had authority to take decisions, whether unilaterally or after consultation with others. Higher proportions of the successful claimed this decision-making authority in the areas of training, maintenance, construction, technical administration, sales, work-study, finance, non-technical administration, production, costing, and finance, than was the case for the sample as a whole. The successful were also more likely to have responsibility for appointing other qualified engineers, they were similarly more likely to appoint office staff, and more likely to have their own secretary.

There appears to be no particular association between success and working in large companies, though there is a connection between success and working in organizations employing a large number of qualified engineers and scientists. There is a strong connection too between success and being in charge of a large number of direct subordinates, suggesting that it is line rather than staff positions which take one into the success league. The successful are also more likely to have authority to spend money without higher authorization: 1,000 DM was the amount specified in this questionnaire item, though it should be added that spending (organizational) money is not such a closely guarded privilege in Germany as in Britain.

The educational achievements of the successful have already been noted, and range from higher average school grades to a higher density of Ph.D.s. More of the successful claimed to be able to speak English and French, and a higher proportion of them travel abroad in connection with their work (this also applies to those with above average degree grades). Along with those engineers with good degrees (or engineering school qualifications) the successful engineers were over-represented among those who had published books or articles (see Table 5.8).

TABLE 5.8

Published books or articles	Whole Sample per cent	Successful Engineers per cent	Good Degrees per cent
	30	44	55

The successful engineers were also prominent in the ranks of those who had given lectures or papers outside their place of work (see Table 5.9).

TABLE 5.9

	Whole Sample per cent	Successful Engineers per cent	Good Degrees per cent
Lectures or papers outside place of work	36	53	54

We should add that academic activity of this kind appears to be much more common among German engineers than among their British colleagues, and holds for graduates and non-graduates alike (this difference is discussed further in Chapter 10).

Perhaps, however, the most significant performance factor differentiating the successful from the sample as a whole and even from those with good degrees (or engineering school qualifications) concerns their showing on patented inventions. By way of introduction it may be remarked that patenting is a more widespread and more seriously regarded activity in Germany. German engineers in general are more likely to have

patented inventions to their credit than are their British colleagues, and this holds for both graduates and non-graduates
(again see Chapter 10). The proportions for the whole sample
and for different sub-samples are shown in Table 5.10.

TABLE 5.10

	Whole Sample per cent	Successful Engineers per cent	Unsuccessful Engineers per cent	Good Degrees per cent	Sub-average Degrees per cent
Patented inventions	24	37	13	31	21

The significance of this finding is that it demonstrates the
importance attached in Germany to the commercial value of
engineering achievement.

Some differences also emerged in the reading habits and
recreational interests of the successful. The successful were less
interested in photography than the sample as a whole, watched
less TV, and were monumentally less enthusiastic about
gardening! They were not distinguished by a greater zeal for
technical reading, but for general reading. This also emerged in
the newspaper and periodical reading habits of the successful.
The most popular newspaper with all the engineers was the
Frankfurter Allgemeine Zeitung (FAZ), a good quality-
national daily. FAZ, however, was even more popular with the
successful. It should perhaps be added that this paper is
reckoned to be good on economic news *and* its Saturday
edition includes an executive job supplement. The quality
weekly *Die Zeit* was also more popular with the successful
though again it was the most popular weekly for the whole
sample as well. The successful emerged as more enthusiastic
readers of *Der Spiegel*, the critical and rather anti-
establishment current affairs weekly magazine founded by
Rudolf Augstein after the Second World War, and of *Capital*, a
business weekly.

In their wider allegiances the successful engineers were not
distinguished by their membership of the *Verein Deutscher
Ingenieure* (VDI), the German engineers association (see
Chapter 6). The successful were, however, much less likely to

belong to a trade union than the sample as a whole, and more likely to belong to the *Union der Leitenden Angestellten,* a representative organization for senior managers. The successful were more likely than the sample as a whole to intend to vote for the CDU-CSU at the next general election. It is also the case for the sample as a whole that high remuneration is associated with CDU-CSU voting.

Another point of interest concerning the successful is what might be termed their successful materialism in private life. Home ownership rates are much lower in Germany than in Britain but the engineers have achieved a British rate of owner-occupancy. The successful engineers, however, were even more likely to be home owners. The survey included an omnibus question on various material possessions, and the successful engineers were in the top place on almost all of these items. Not only were they the most likely of all the sub-groups to own their own homes, they were also in the lead on second cars, colour TV, the ownership of boats, and foreign holidays. The 'runners-up' on most of these items, incidentally, were the engineers with good degrees or engineering-school qualification grades. It is worth repeating that the showing of the successful engineers on these possessions and expenditures is not entailed by the definition of success. The successful are not absolutely the richest, just the richest for their age-group, and it is a point of interest that their incomes are used in these ways.

We referred earlier to the fact that the successful engineers were distinguished by their retrospective evaluation of their college experience, rating it more favourably than the sample as a whole. The successful exhibit a similarly positive attitude to the job and to their occupational status; they view these, that is, more favourably than the sample as a whole. The group of successful engineers exhibited a higher level of job satisfaction than their colleagues, in particular it being noticeable that a higher proportion of the successful replied to a questionnaire item on sources of dissatisfaction with the answer 'I am satisfied in every respect'. The successful showed a similarly high level of satisfaction with the general status of the engineer in German society. Furthermore, in response to a questionnaire item inviting respondents to place ten occupations, including that of engineering, in a prestige order according to their personal view and the estimation they thought the general public held of these jobs, the successful placed engineering higher in the list than the sample as a whole, and depicted the general public as doing so. Finally, in this connection, a higher proportion of the successful thought that their non-technical colleagues held a favourable view of them as engineers.

Relatively little has been said so far about the mechanisms whereby success is attained. Our view is that to a large extent the characterization of the successful is itself a sufficient explanation. They come, on average, from a more favoured social background in terms of both the occupational status and educational level of their parents. In addition, they are early and resolute in their choice of engineering as a career, and are consistent over-achievers at school, at college, at work, and even in terms of foreign language ability and technical publications. They lead the whole sample on patented inventions, and showed greater breadth than their colleagues in their choice of subjects, as young men at college, and show it still in their reading habits. They have a general preference for work in industry rather than in the public sector, and substantially greater responsibilities than their colleagues. Their attitudes reflect their achievement and personal security, and they are proud of being engineers.

There are, however, perhaps three points relating to the mechanisms of success which deserve extra comment. First, although private schools exist in Germany they do not have the same place in German life as the public schools in Britain have in this country. The German private schools, that is, do not see themselves as training an élite, no particular status attaches to attending them as opposed to the state *Gymnasium*, and they are not favoured especially by the German upper middle class in general. All this represents a contrast with Britain, and even a contrast with regard to the dynamics of the career success of engineers.[3]

The second point concerns any possible relationship between job mobility on the one hand and higher income and/or success on the other. There appears to be some kind of a positive relationship between job mobility and higher earnings in general, but this relationship is neither strong, nor unambiguous. To cut through the complexities the data suggest that one or two job changes up to a certain age or within an age range probably lead to higher average earnings: the data do not suggest a progressive 'straight-line' relation between mobility and remuneration. Between mobility and success as defined in the present chapter there appears to be no relationship at all. The only relevant connection to emerge from the analysis of the data is a correlation between non-mobility and failure (in terms of being in the bottom quartile of the salary range for the appropriate age group). In other words it may be disadvantageous to stay put when there is a good reason to move, but nothing is to be gained by moving for moving's sake.

Third, the survey asked the engineers to give their reasons for the various job changes which they made. Now in the response to this there are naturally some elements which are common to the sample as a whole, but the differences are instructive. The successful engineers were relatively less interested in immediate salary increases, more technical work, better working conditions, better general atmosphere, and in finding a nicer place to live. What did interest them, and this is the biggest single difference between the successful and their colleagues, was promotion prospects. This is not a platitude. The successful may have greatness thrust upon them, or they may actively pursue it.

It was suggested towards the beginning of this chapter that the popular German claim to have evolved a classless society is not substantiated by the data on the social background of the successful and their less successful colleagues. On the other hand a general characterization of the successful engineers in the sample does at many points lend support to the view that German society is the *Leistungsgesellschaft* (achievement society) which it so often claims to be. In the context of this second judgement the views of the engineers on their social class membership are of interest. In response to a question inviting them to label themselves in social class terms 93 per cent classified themselves as middle-middle class, or better, although only about a third of the sample were born to this class. In view of this self-flattering classification the interesting cut-off point is in fact that between those who see themselves as (only) middle-middle class and those who place themselves in the upper-middle-class or upper-class categories (see Table 5.11).

TABLE 5.11

Self-assigned class	Whole Sample per cent	Successful Engineers per cent	Unsuccessful Engineers per cent
Upper class	3	6	2
Upper middle class	30	43	17

There is something compelling about German class theory: if you earn more than three-quarters of your own age-group do, you must be upper middle-class.

Chapter 6

The Engineer and his Professional
Institution

There are historical and social reasons for the considerable differences that are found between the constitutional positions of engineering institutions in Britain and in Germany. In Germany long ago the initiative for educating and training professional engineers came from the state, whereas in Britain engineering education gradually developed from craft-training. Not until the mid-nineteenth century in Britain did engineering become respectable and accepted as a university subject. The result is that an engineering qualification in Germany, either Dipl. Ing. or Ing. Grad., is not only highly respected but is legally protected and recognized as a full professional qualification. In Britain, however, a university engineering degree is not so highly regarded and does not *per se* give full professional status. Professional qualifications are not awarded by the state but by various engineering institutions, the oldest of which, the Institution of Civil Engineers was founded in 1818. The Institution of Mechanical Engineers, the next oldest, was founded in 1847 by a famous railway engineer George Stephenson, largely because the Civil Engineers would not accept him as a member.

Nowadays to achieve professional status a mechanical engineer must also satisfy requirements for industrial training and experience laid down by the Institution of Mechanical Engineers, pay an entrance fee, and thereafter pay an annual subscription depending on his grade, either of Member (M. I. Mech. E.) or, if he is particularly senior, Fellow (F. I. Mech. E.). No other European country has a system of professional qualifications controlled and awarded by private engineering institutions. In Britain there are now sixteen engineering institutions, each of them set up independently as the profession developed and became more fragmented. Each of these has a royal charter empowering it to regulate industrial training in its particular branch of engineering and to award professional engineering status to its members. These institutions are specifically precluded by their charitable status which earns tax rebates, from doing anything that would be of direct financial benefit to their members. They cannot therefore operate as trade unions and have little or no influence on salaries or terms of service.

Collectively the institutions formed a Chartered Institution in 1965 called the Council of Engineering Institutions from which it was hoped that a single engineering institution might grow, but there is little sign of this happening. The two largest institutions are the Institution of Electrical Engineers and the Institution of Mechanical Engineers, each with a corporate membership of about 74,000. It is possible that this system for awarding and controlling professional qualifications may soon be changed as a result of recommendations of the government committee set up under the chairmanship of Sir Montague Finniston[1] to look into the engineering profession, and that Britain may move towards a state-regulated system closer to that of other European countries. However at the moment Britain is different.

On the mainland of Europe a new graduate with an engineering degree from a state university is entitled to immediate professional recognition. His five or six year course, the last two having a pronounced industrial slant, together with an obligatory period of industrial training (before or during the course) gives him full professional status. In Britain, on the other hand the new graduate after his three year university course must undergo at least two years of recognized industrial training before he may apply for corporate membership of a professional engineering institution. Nevertheless the lack of professional status does not, at the moment, prevent a British graduate from practising in most engineering jobs, although there are a few exceptions, such as gas or mining engineering, where there are statutory requirements for obligatory practical experience. Most employers do not insist on corporate membership of the Institution of Mechanical Engineers but usually require an engineering degree. Partly as a result of this and partly because young engineers feel that institutions have little to offer, only about half the current graduates join an institution, although this does not seem to jeopordize their careers.

In Germany the largest engineering institution is the *Verein Deutscher Ingenieure* (VDI) founded in 1856 and originally open to everybody concerned with *Technik*, the art of manufacture. Its history is well summarized by Hortleder[2] who points out that in its first thirty years a large proportion of the members were entrepreneurs and owners of firms. Many of these were both engineers and entrepreneurs but the proportion of this type dwindled because of the predominating engineering interest and by the turn of the century about half the members were engineers. By now its membership almost entirely comprises engineers, but although most of these are practising engineers the members of the Council of VDI are mainly directors and therefore represent the top management of firms. In this sense the Council has now become an élite managers' club and a pressure group rather than furthering *Technik* and the well-being of German industry which was its original purpose, though this judgement does not apply to the rank and file membership or the normal activities of the institution.

VDI covers all branches of engineering but mechanical engineers now form the majority of its members. Nevertheless only about 20 per cent of all engineers in Germany, both Dipl. Ing. and Ing. Grad., are members. This is partly because there are smaller and more specialist institutions such as the *Verband Deutscher Elektrotechnik* (VDE founded in 1893) for electrical and electronic engineers. The VDI does have a kind of legal charter, but not in the British sense, to act as a qualifying body. Most of the members are Dipl. Ing. and Ing. Grad. in proportions similar to their distribution in Germany, and only a very few members are not so qualified. It is therefore more of an engineers' association than an institution in the strict British sense and functions mainly as a forum for meetings, conferences, and special post-experience courses, some of which form TV series. A large part of its income comes from its activities as a technical publishing company, the output of which includes excellent engineering guides and codes of practice which are the result of the work of over 500 technical committees. It also commissions surveys and reports on important long term trends, such as the future demand for engineers, carried out by the Batelle Institute but sponsored by

VDI. Like the British Institution it has no trades union function.

We were advised, when planning the survey, that VDI members would show characteristics different from those of the sample average, and would stand out as an élite group, and this was true in some but not all senses. The VDI members were not differentiated by inherited or present social class distinctions; there was no evidence that they were better paid than engineers in general; and their schooling, recruitment, and training followed very similar patterns to those of non VDI members, but there was a slightly above-average proportion of Catholics. But they were more likely to have attained a good degree, whether it was Dipl. Ing. or Ing. Grad., both of which were almost equally likely to join VDI. VDI members also tended to be older than the sample average, 45 per cent of them being over forty-five, compared with 28 per cent of the whole sample.

The VDI members, in line with their superior degree grades, were also more likely than the sample average to have published books and articles, and to have given technical lectures outside their place of work.

Furthermore there were some small indications of greater job satisfaction and professional loyalty; 85 per cent said that they would choose engineering as a career if they were starting again, compared with an 82-per-cent sample average, and the VDI members were considerably more likely to read technical journals regularly than were non-VDI members. Their children were under-represented in the specialist technical and economics grammar schools, but over-represented at the traditional grammar schools.

Thus it was not possible to substantiate in all respects the élite characterization of VDI members that a number of German advisers had suggested, although it is clear that their Council members are among the top management élite of German industry, and the rank-and-file members an élite of technical interest and engagement. It is difficult to know what prompts an engineer to join a professional institution, other than it was 'the thing to do' in the 1920s and 30s when many of the VDI members graduated; nevertheless we do not know when the members now in their fifties did in fact join. They may only have joined when they emerged as successful engineers. There is less tendency among the present generation to join VDI, but this may be due to social changes rather than age differences. The one anomalous response of VDI members in the survey was that, although they were more satisfied with their job, which they regard as a contribution to the community, they were less satisfied with the general status of the engineer in modern society. This could again be explained by VDI members being in the older age group and feeling that their status was not as high as it used to be, whether this feeling is justified or not.

Finally, it should be said that although we have spoken here of 'professional institutions' this phrasing is not really in line with German usage. It was hinted in the Introduction that the Germans do not really have a concept of the profession; they do not that is, single out some occupations, label them as 'professions' and endow them thereby with an additional element of prestige. The implications of this for the standing of engineering in particular are developed in some detail later (see Chapter 9).

There is, however, a paradox here. If profession is the opposite of non-profession, and one cannot make the distinction in German, professional is also the opposite of amateur. The Germans, with their technical virtuosity and national emphasis on vocational education do not have many amateurs. And the paradox is that one can really only give expression to these ideas in a book in English.

Chapter 7

The Engineer Off Duty

If one wishes to have a sense of the complexity of the modern world the essays of the German sociologist Georg Simmel provide a good starting-point. This writer was fascinated by certain social types — the stranger, the nobleman, the beggar, the adventurer and so on — and he produced what are generally agreed to be perceptive accounts of these roles.[1] But his subjects were part of a completely different social pattern from that of today, as can be seen by his one dimensional treatment of them. In its sociological context being a nobleman is a full time job, and there is no question of appraising him in different settings. Yet in the modern world one's job, however absorbing, does not account for every hour of the day, there is home and leisure as well, and the various interactions between all these spheres of life. In this chapter we would like to add another dimension to the portrait of the German engineer by offering some account of his life away from work.

A discussion of the German engineer off-duty provides both a more complete picture of him and gives clues for a fuller understanding of his integration in German society. At the same time this analysis yields one or two 'unprogrammed insights' and also opens up the question of the way the engineering profession is going to perpetuate itself in the next generation. Looking at the engineer in the non-work context affords an opportunity to check out some preconceptions, and in particular, those about his social and political attitudes.

If we begin by posing the seemingly simple question, are engineers sociable? It has to be conceded that there are two different answers. If one looks at the responses relating to joining voluntary organizations, it would appear that German engineers are most certainly unsociable, but the answers to other questionnaire items make it abundantly clear that they are by no means recluses or socially isolated. In fact 77 per cent of the sample had some purposeful social contact with close relatives (other than those with whom they lived) in the month preceding the interview. Another 62 per cent of the sample similarly claimed some out of hours social contact with their immediate work colleagues in the same period, and another 41 per cent referred to social contacts out of work with other friends from their place of work who were not *direct* work colleagues. A further 47 per cent referred to social contact with neighbours in the same period (the wording of the relevant questionnaire item was strong and certainly precluded just 'passing the time of day'), and 73 per cent of the whole sample claimed genuine social contact with friends not coming in any one of the previous relative–work colleague–neighbour categories.

This non-recluse interpretation is supported by two other sets of data. The first is the clearly zestful attitude to holidays. Only 2 per cent of the sample declined to answer a question concerning the holiday they had taken in the previous year, and conceivably did not take one. Of those responding to the question only 28 per cent had taken their main holiday in Germany itself, the remaining 70 per cent having been abroad. For this majority holidaying abroad the most popular countries, in descending order were:

Austria
Italy
Spain
Scandinavia
France
Switzerland

Britain was in tenth place after the above six, followed by Yugoslavia, Holland, and North Africa. This relative unpopularity of Britain as a holiday destination is surprising in view of the fact that English is the foreign language that German engineers, and Germans in general for that matter, are most likely to know. Also, it is fair to add, the price differences in the mid 1970s should have made Britain a 'looter's dream' for well-paid continental engineers.

The second set of sociability-confirming data concerns quite simply — friends. One of the questionnaire items asked how many people the individual engineers counted as their close friends. Only 5 per cent of the whole sample did not claim any close friends (bear in mind that this is an all-age sample where the older members may well have been more involved with family and grandchildren). Thus 95 per cent claimed various numbers of close friends, with 32 per cent putting the number at 5–9, and 22 per cent putting it as high as 10–19. This again does not suggest that the engineer is any kind of recluse, only able to commune with his machines.

This question of friendship was pursued in the survey where subsequent questionnaire items asked the engineers to give the occupations of those they claimed as close friends. One interesting thing that comes out of this data is evidence of cross-class friendships. It is certainly the case in Germany that engineering enjoys higher standing than in Britain (this proposition will be argued in some detail in Chapter 9), and in terms of educational level and income as well, German engineers are regarded as solidly middle-class. The interesting thing is that a lot of those they claimed as close friends were more like working class in terms of their jobs. These manual occupations from which such close friends of the engineers in the sample came tended to be skilled, craft, or technician: in fact 145 engineers named technicians as among their close friends, 60 named skilled workers, 20 electricians, 19 foremen, and 13 fitters.

These cross-class friendships are a small reflection of the differing nature of the German class system. Although in Germany objective differences of income, job status, and educational level exist as elsewhere in the industrialized world, the complex of behavioural-stylistic-attitudinal differences with which we British are familiar are much less marked in Germany. Thus a close friendship between some member of a solidly middle-class occupational group — and engineering is certainly that in Germany — and, say, an electrician, is that much less remarkable: any gap to be bridged is smaller.

In fact the top six occupations for engineers' friends were as shown in Table 7.1.

This list contains another point of interest. That engineers are a top choice as friends is no surprise, but the sales managers in second place are surprising. These sales managers, in the German scheme of things, will mostly be qualified in business economics (there are both graduate and non-graduate qualifications in business economics corresponding to the Dipl. Ing.

TABLE 7.1

Occupation	Number
Engineers	644
Sales managers	260
Teachers	148
Technicians	145
Doctors and dentists	102
Officials	81

and Ing. Grad. in engineering), so they clearly do not share a common educational background. What is more the engineers claiming these close friendships with sales managers will not for the most part be working in Sales themselves but in R and D, Design, Production, Testing, and so on. In the main line technical and production functions, that is, that are often depicted as having little natural rapport with Sales.

In a different context one of the present authors has spent a lot of time as an observer in German manufacturing companies, and noted two things possibly relevant to this engineer—sales manager link which has turned up in the survey. One is that management division of labour seems less sharp than in either corresponding firms in Britain or in the (exhortatory) American management literature. The second is that the tendency to see Sales as a technical, or technically shaded exercise, is stronger in Germany.

The hobbies and interests which appealed to the engineers in the sample were numerous, but some of their leisure pursuits were mentioned by large numbers. The top ten, in fact, were as shown in Table 7.2.

TABLE 7.2

Sport
Gardening
Reading
Walking
Technical reading
Music
Photography
Craftwork
Job-related hobbies /travel
Learning foreign languages

Some of the little sub-sample variations are interesting. The Dipl. Ings. were enthusiastic about general reading, and the Ing. Grads. about technical reading. The successful were less interested in photography than the unsuccessful (in the senses defined in Chapter 5) as well as being much less enthusiastic about gardening, as was noted earlier. It is only a small point but photography does not seem to have the same status as an acceptable manager interest as it does in Britain. Engineers working in industry were much more interested in learning foreign languages than their colleagues in the public sector, and this probably represents a perceived occupational need. In our experience there is a tendency in German firms to regard sales, especially export sales, as everyone's business. Another small point of interest is that craftwork hobbies were almost equally popular with Dipl. Ings. and Ing. Grads., although the training of the latter is more practically oriented. What this probably reflects is the German view that making things is worth doing and not beneath anyone's dignity no matter how augustly qualified.

They are not averse to spending time on domestic pottering. One of the questionnaire items asked how much time they spent on jobs in the house, gardening, or working on their cars in what they would view as a normal week, and the answers are shown in Table 7.3

As with the constellation of hobby and leisure interests there were some smallish differences between the sub-samples. It is the trade union members, the unsuccessful, those with below-average degree or engineering school qualification grades, and

TABLE 7.3

Hours per week spent on work in house, garden, and car	percentage of sample
None	2
1–4 hours	24
5–9 hours	30
10–19 hours	32
20–29 hours	7
More than 30 hours	2
Not answered	3

those working in the public sector who emerged as most 'home-centred', with their opposites in all these cases spending a less than average time on these activities.

The sample as a whole were more moderate with regard to television-watching. The engineers were asked to compute how much TV they had watched in the seven days before the interview, and the distribution is shown in Table 7.4.

TABLE 7.4

Time spent watching TV over 7 day period	percentage of sample
Up to 5 hours	47
6–9 hours	23
10–14 hours	22
15–19 hours	3
20 hours and more	4
Not at all	6

Thus in round figures about half the sample were averaging an hour a day viewing or a little less. The Ing. Grads. watched more television than the Dipl. Ings., but it was the trade-union members who emerged as the 'heavy viewers'.

If we return to the questionnaire items on joining voluntary associations, other than sports clubs, it will be seen that the German engineers were definitely not 'joiners'; only 4 per cent of the whole sample belonged to any organization. Their distribution is shown in Table 7.5.

TABLE 7.5

Type of Voluntary association	No. of members
	(Sample size 1,006)
Automobile clubs	13
Student Associations	12
Political Parties	12
Cultural Activity Associations	9
Church organizations	3
Military Associations	3
Charitable Organizations	2

These absolute numbers are so small that it is not possible to observe any trends among the sub-samples.

There were two exceptions to this shyness where clubs and organizations were concerned, though they are both work-related and cannot strictly be termed 'off-duty'. The first was that almost 30 per cent belonged to VDI, and were in fact over-represented compared with the national proportion of German engineers joining a professional association (see Chapter 6). The second exception was the number who were trade union members. Unfortunately it is not possible to say how many or what proportion of German engineers join trade unions. The authors made extensive inquiries on this point in the early stages of the study and are convinced that no one knows. It would appear that neither the unions concerned, nor the DGB (*Deutscher Gewerkschaftsbund,* the German equivalent of the TUC in Britain), necessarily have breakdowns of their membership according to job and qualifications. If this seems surprising it should be added that the German unions are industrial unions open to *all* employees from a given industry, irrespective of trade or skill level. Thus they are much more heterogeneous in respect of the job, skill, and qualifications of their members than are many British unions. To put it into perspective, for Germany's largest trade union the IG Metall, to make an exhaustive catalogue of such attributes of its members would be an undertaking equal in scope to a national population census in Switzerland.

Of engineers in our sample 15 per cent were trade union members, and this may approximate to the national proportion though one cannot be sure of this. This minority claiming trade union membership was spread over several unions, as indicated in Table 7.6.

A point of interest which emerges from the engineer 'off-duty' is that there are several occasions where the responses from the sub-samples of those with good degrees, those deemed occupationally successful, and those working in industry seem to overlap considerably, usually in the direction of having less time (or inclination) for homely pursuits. This combination of sub-samples turns up again in the education and job aspirations of the children of the engineers.

In constructing the questionnaire our feeling was that to

TABLE 7.6

Trade Union	percentage of whole Trade Union Membership
I. G. Metall (metal industries)	42.76
Deutsche Angestellten Gewerkschaft (white collar employees)	21.71
Gewerkschaft Öffentliche Dienste Transport und Verkehr (public service)	11.84
Deutsche Postgewerkschaft (post office)	11.18
Gewerkschaft der Eisenbahner Deutschland (railways)	7.89
Other trade unions	4.60

inquire as to the job aspirations of younger children would simply evoke the aspirations their parents held for them. Consequently we asked only about the job aspirations of children over 14. The preferred and anticipated jobs for the first-born 14-plus children of the engineers in the sample, in descending order, were as shown in Table 7.7.

TABLE 7.7

Graduate engineer (i.e. Dipl. Ing.)
Teacher
Doctor, dentist, vet, or chemist
Salesman with degree in business economics
Scientist

The number of all other occupations was small, never more than six.

The preferred jobs of second children of the same age were not very different. Again they are listed in descending order of popularity:

TABLE 7.8

Graduate engineer (Dipl. Ing.) ⎫ joint
Salesman with degree in ⎬ top
business economics ⎭
Teacher
Doctor, dentist, vet, or chemist
Social worker

We noted earlier that sales managers came up quite often in the ranks of the close friends of the engineers in the sample. It is again interesting to see that the two occupations were linked as alternatives in the minds of the engineers' children. As with the choices of the first-born children, the numbers for all occupations other than those listed above were small, not more than six apiece. The absolute number of third-born children in the sample was too small to permit much analysis but again the first choice was graduate engineer.

There are two particular things which should be underlined in this review of the occupational choices of the engineers' children. The first is the sheer overwhelming popularity of engineering. In this connection it should be kept in mind that the reference is to the engineers' children, not to their sons. Female engineers are as thin on the ground in Germany as they are in Britain, so it will be clear that engineering as an occupation was highly popular with the engineers' sons.

The second thing is that once again the same combination of sub-groups emerges; that is, it was for children of the succesful, of those with good degrees, and of engineers working in industry, who were more likely to name engineering as their future career. There is a further piece of evidence which should be cited. Besides the various kinds of traditional *Gymnasium* (the German equivalent of the grammar school, see Chapter 2) there exist specialist technical and economics grammar schools. Attending these specialist schools naturally tends to imply reading engineering or economics at university and the likelihood of an eventual career in industry. Of the engineers' children who were undergoing secondary education at the time of the survey, or who had already passed through the secondary phase, a sizeable proportion were attending these specialist grammar schools, a proportion only slightly smaller than that attending the various traditional types of *Gymnasium*. And again there were three cases in which the proportion attending these technical and economics grammar schools was above average for the sample as a whole: the children of the succesful engineers, those with good degrees, and those working in industry.

These particular findings are of interest because we can make reasonable inferences from them concerning the standing of engineers and the status of industry, themes which will be explored in more detail in the two following chapters. This juxtaposition of sub-groups is interesting in another way. In studies of occupational groups this endogamy often comes to light, where, for example, most doctors had fathers who were doctors. There is also a tendency to regard such a finding (of occupational endogamy) as self-explanatory. If, however, there is a wider moral to be derived from this part of the present study, it is that a differential mechanism of inter-generational influence and emulation may well be at work.

It was noted earlier in the present chapter that relatively few engineers were members of voluntary associations. This finding is not, as has been shown, consistent with other evidence of sociability and social integration, but it might be thought to be consistent with the engineers' presumed naïveté or apoliticality. The idea here, and it is traditional and diffuse rather than attributable to particular sources, is that engineers are more at home with their machines and the dynamics of the inanimate than they are with social and political processes. Thus they are not especially informed about the latter and certainly not socially and politically active (viz. low membership of all kinds of voluntary associations). Metaphorically speaking, engineers have had a 'bad press' with regard to their putative social and political attitudes. To put it bluntly, there is a tendency to attribute to them a strain towards right-wing conformity and reactionary traditionalism.

Against this misty background of assumption and ascription we now have a very good, and recent, study of the social and political attitudes of German engineers. It is a study which throws considerable doubt on both propositions, namely that the engineer is an uninformed and apolitical member of the community, and that he tends towards right wing and traditional attitudes. The study in the form principally of a very large survey was the work of a former German professor of politics, Eugon Kogon.[2]

Eugon Kogon is probably best known in Germany on account of his experiences during the Third Reich. He fell foul of the Nazi regime and was imprisoned in Buchenwald concentration camp from 1937 until the liberation of the camp by American paratroopers in April 1945. These experiences gave

rise to one of the first books about the concentration camps –
Der SS Staat (American edition entitled *The Theory and
Practice of Hell*, 1950). This work is not a purely biographical
account; Kogon sought to produce a general account of the
concentration camps and the SS, supported in part by his
personal experiences. *Die Stunde der Ingenieure* (*The Hour of
the Engineers*) is the book in which Kogon reports his survey of
German engineers and it is the first he has published since *Der
SS Staat*. There is some connection between these books.

Die Bewältigung der Vergangenheit (the conquest of the
past) is a standard phrase in Germany. Much post-war
German fiction, and some social science research, is concerned
with the National Socialist experience, the problem of guilt,
and the task of moral reconstruction. A basic question for this
genre is: how was national socialism possible? In this allocation of blame the engineers have a significant, if not central,
role. It is not suggested that they actively aided the rise of the
Nazis, as for instance did some leading industrialists, nor that
they facilitated this rise by inactivity, as did the military
establishment. The indictment is rather one of passive com-
pliance. The engineers were uncommitted and apolitical; they
allowed their knowledge and talents to be used. Gerd Hortleder in his historical study of the VDI[3] quotes from Albert
Speer, Hitler's armaments minister and an architect (in the
German scheme of things an architect is an honorary engineer!), on the title page: 'One simply got on with one's own
business, and as far as possible did not bother about what else
was happening'.

This is the phenomenon which has concerned Kogon; and a
third of a century later he has made an empirical study of the
socio-political attitudes and engagement of German engineers.
The core of his study is a questionnaire survey of a sample of
over 25,000 engineers and scientists (engineers constitute the
majority) in which they were quizzed in detail about their level
and sources of political information and about their views on
certain social and political questions. There are two important
general findings to emerge from this study.

The first is that Kogon, after a detailed questioning of his
sample on their reading, including newspapers and journal
reading, and their TV viewing, with especial reference to cur-

rent affairs programmes, concluded that there was no evidence for the view that engineers lack interest in public and political affairs, or of a lack of information about them. The second general finding was that engineers turn out to be not at all right-wing, traditionalist, conformist, or submissive to authority, at any rate in their responses to the items on Kogon's questionnaire. It is worth illustrating this. Much of the questionnaire was in the form of (contentious) propositions on social and political questions with which the respondents were invited to agree or disagree. They were also asked some direct questions about their position on the political spectrum, the development of their political consciousness, and so on. The following are among the findings from this part of the study.

The majority of the engineers in Kogon's sample regarded their own education as too one-sidedly technical. Their early political memories were dominated by the traumas of the Weimar Republic, the Nazis, the War, and the division of Germany. They were sympathetic to the idea of a United States of Europe, but hostile to the extra-parliamentary opposition. They were in favour of the United Nations but pessimistic about UNO's ability to maintain world peace. The majority of the sample were not in favour of more national conciousness in Germany. They approved the recognition of the Oder Neisse line, and of the German Democratic Republic. They avowed themselves in favour of co-determination in industry, supported the trade unions, and were even in favour of pupil co-determination in secondary schools. They were critical of the concentration of economic power in Germany, and in favour of greater public control.

About half of the engineers were worried by society's loss of ideals and standards, a significant minority were worried by growing sexual freedom, and a small minority were in favour of corporal punishment. If these last sound more in line with the views more popularly attributed to engineers it should be added that the majority took a strong line on the moral implications of technology and individual responsibility for its utilization. They were decidedly lukewarm on military values, but in favour of equality of opportunity for women, and opposed to a class-determinist view of intellectual and creative abilities.

On the subject of self-selected political labels the two most popular political epithets were 'liberal' and 'anti-communist'. Only 10 per cent described themselves as 'conservative' and a higher proportion labelled themselves socialist. In response to a separate question the majority placed themselves on the left half of a finely differentiated political spectrum.

This quick itemization of some of Kogon's findings should not be viewed as a comprehensive summary of his extensive study. On the other hand we have not selected from the study in a one-sided way: this liberal responsible stance is precisely what emerges from the survey as a whole.

In view of Kogon's detailed and influential study we may end this review of some aspects of the German engineer's private life with a speculation. It is a theme of this book that there are sufficient differences in the situation of engineers in Britain and Germany to render a study of German engineers interesting in Britain. When it comes to this particular thesis concerning the engineer's socio-political attitudes, however, and in the absence of comparable studies outside Germany, it may well be that we should assume engineers in general to be more liberal and progressive than has generally been thought. At least we should do so unless or until some significant counter-evidence is produced, whether in Germany or elsewhere.

Chapter 8

The Status of Industry

In recent years the concept of the status of industry has gained currency and credibility. It has often featured in discussions of the British economic performance, the argument being that the status of industry is not high in Britain, and not as high as in several other, and richer, countries. This fact, it is suggested, has, in Britain, negative consequences for recruitment, morale, and other intangibles in the national life.

Our view is that this argument has some force, although there is no suggestion that the status of industry alone determines national economic performance. Furthermore West Germany is certainly one of those countries in which industry does enjoy a higher standing, this does have positive implications for the way the economy functions, and it is a very significant factor in any estimation of the status of German engineers as well.

The purpose in this chapter is not primarily to seek to prove the proposition that the status of industry is high in Germany. It is a diffuse proposition which is difficult to prove (or disprove) in a situation where there is no agreement about what would actually count as evidence. The present intention is rather to interpret this aspect of German society: to suggest reasons for this presumptively higher status enjoyed by German industry, to draw attention to some of its manifestations, and finally to work out the thesis with regard to the engineers in our survey. It is possible to begin this exercise in a quite homely way!

For anyone who spends time in Germany and is used to hearing people talk about work and job choices it soon becomes clear that industry is differently perceived. Certainly 'going into industry' is regarded as a stronger career choice, a natural choice for the able and ambitious, and certainly not one requiring any apologetic rationalization. The German equivalent of 'he's just come down from Balliol with a first in Greats and is going to the Foreign Office' is something like 'he got his Dipl. Ing. in nine semesters at Aachen and is going to Siemens'.[1]

Neither does there seem to be any feeling in Germany that a career in industry is not quite the thing for a gentleman. In saying this it has to be conceded that there is no way of knowing what the Germans understand by gentleman. They do not have a word in German for gentleman, but use the English word.

It may be that German industry is less encumbered by folk-memories of the early phase of industrialization and the rapid urbanization which accompanied it. Say 'industry' to an English schoolboy and he will probably at once think of ill-treated child labour in a Lancashire cotton-mill 150 years ago, or some comparable atrocity. The German schoolboy is more likely to think of some industry or plant he has visited, or, if he has reached that level of maturity, of the overall economic performance. It is only a small point of contrast but in Britain industrial visits for schools more usually involve the lower forms in the academic hierarchy and, before the advent of comprehensive schools, groups from secondary modern rather than grammar schools. In Germany, on the other hand, visits to firms by groups from the *Gymnasium* (grammar school) are quite normal, as are visits to public utilities such as power stations or gasworks, offering some technical interest.

It is interesting that the German folk-memory does not seem to return so readily to whatever evils accompanied the birth of industrialism in that country. This may reflect a disparity in historical assessment. For British people reviewing the historic past these evils may appear as the worst which ever befell Britain. For Germans the worst thing that ever happened occurred between 1933 and 1945 and has nothing to do with industrialism. Or to put it another way, Germans probably have difficulty in bridging this 1933–45 gap; it is as though the origins of the world they live in may be traced back to 1945, but everything before that belongs to 'another country' and another age. It is different for us British, with our centuries of peaceful change and flair for continuity.

Another difference between the two countries which impinges on the perception of industry is political. This difference is probably more apparent than real, but appearances have their meaning. In Britain the Conservative Party is thought to be 'for industry', but there are doubts about the Labour Party. The latter, of course, wants prosperity for the nation and the reduction of unemployment but there is the vague feeling that industry is just tolerated as a means to the ends of prosperity and full employment, neither loved nor esteemed for itself. No such 'vague feeling' exists in Germany. There all the main parties are positive in their attitude to industry, and the SPD (German 'labour party') unequivocally accepted capitalism, free enterprise, and private industry in their Bad Godesberg platform in the early 1960s.

It is probably also fair to say that industry in Germany, from the point of view of, say, the young graduate contemplating his career options, lacks some general off-putting features in comparison with industry in Britain. The most obvious thing, of course, is the difference in industrial relations. Strikes are much rarer in Germany, so a major uncertainty is, relatively speaking, removed. Add to this the facts that wild-cat strikes in particular are very unlikely, because of the legalistic nature of German wage agreements and compliant attitudes to the law, and demarcation disputes are virtually impossible because of an industrial relations system dominated by a small number of large industrial unions, and it will be clear that the life of a German company, relatively free from these kinds of disruption and antagonism, proceeds on a more even keel. Our putative young graduate, contemplating the advantages and disadvantages of a career in German industry, is not likely to be worried by fears of confrontations with workers or their representatives. From this point of view a university lecturer in Germany has a tougher job.

This relative peacefulness appears on a wider front in German firms than simple employer–employee relations. One of the present authors has spent a lot of time as an observer in various German companies, where a major impression is the fairly orderly, stable climate. There simply appear to be fewer crises, shortages, shut-downs, and changes of direction. All this means that there is less call for the German manager to be a trouble-shooter, crisis-handler, instant reactor, and so on. Or to put it another way, the transition from a full-time education course to a German factory is not so disjunctive as in Britain. It is also noticeable that German managers do not indulge so much in jungle allusions and military metaphors.

Again, what might be termed a cumulative social effect is in operation in German industry. German firms employ more people with higher formal qualifications, especially technical qualifications. There are more university graduates, more Ph.D.s, more people with the *graduiert* qualification (see Chapter 2). Furthermore, these trends reach further down the size scale, so that relatively small firms by British standards will be found employing such qualified manpower: this clearly emerges from our survey when the data on sizes of employing organisations are placed alongside comparable data for Britain. The effect, of course, is to render German industry generally more hospitable to the well-qualified: they will constitute less of a minority, and will experience less of an assimilation problem.

This tendency towards the undramatic absorption of the well-qualified is enhanced by a factor mentioned in the previous chapter in the discussion of German engineers friendship patterns. This is the less intrusive nature of social class differences in Germany. Whatever the objective facts of income or educational differences, the behavioural and stylistic differences seem much less in Germany. Thus our putative young graduate, whatever happens to be his own social background, is likely to worry less about whether he will be able to 'deal with' the workers, or relate to 'ordinary people'.

All this is admittedly a dialectic roundabout, with the gaily coloured cars going up and down to the music of a master plan. Industry has higher status, attracts more well-educated manpower, which make it easier to attract still better educated manpower, which raises its status further. Less intrusive class differences (*inter alia*) make for better industrial relations, fewer strikes, more output, greater diffusion of wealth, more working-class affluence, and even fewer class differences.

Earlier reference was made to one historical aspect, the possible 'blocking effect' of the national socialist period in matters of German retrospection. There is perhaps another historical consideration which, while it may not explain the status of industry in Germany, throws some light on a configuration which is part of that status. It is generally acknowledged that Britain's Industrial Revolution came much earlier than Germany's. We do not wish to contest this, but to suggest that it is also of interest to note the timing of the key phase of industrialization in relation to other developments in the two countries considered separately. Let us start with Britain.

When Britain's industrialization began, taking the conventional 1760-plus dating, the state of science and engineering was not advanced, and the stock of formal scientific and technical knowledge was relatively slight. Thus it was not to be expected that this early phase of industrialization would be pioneered by graduate engineers and Ph.D.s in chemistry, and it was in fact pioneered by a miscellany of gifted craftsmen, interested amateurs, and technically sensitive entrepreneurs.

The second relevant consideration is that we could scarcely speak of an education *system* in Britain in this period. At the time Hargreaves invented the Spinning Jenny the law which would provide for compulsory primary education was more than a hundred years off, and the legislation which would inaugurate a standard system of secondary education was nearly 150 years away. There were only two universities in England though Scotland was better endowed in this respect. In short it was not reasonable to expect the world of education to fuel the early phase of industrialization, with personnel or ideas. Nor, conversely, was it reasonable to expect the world of education to react positively to the needs of industry at this stage.

It should be underlined that this is advanced as a general argument. Obviously there were some instances of inputs from the world of learning and of reactions in it. The rise of the Dissenting Academies with their emphasis on science and foreign languages is one example; the eventual emergence of the Mechanics Institutes is another. But, we are asserting, this did not and could not happen on a wide front for reasons of relative timing.

The third simple thing about the British Industrial Revolution is that it succeeded. What we may have missed is the seductive effect of this success. If industrialization could be launched without technically qualified manpower or a positive relationship with a relevant and responding education system, why should anyone think these would ever be necessary? This is the contrast with Germany.

In Germany industrialization occurred later, against the background of a much more developed education system, and concomitant with advances in the natural sciences and engineering knowledge. These elements formed a natural alliance, and it has persisted. This is clear in many of the features of the German education system today.[2]

If we start at the top, with a view to illustrating this contention, it should be noted that there are eleven technical universities in West Germany, having parity of esteem with the traditional universities, and dating mostly from the nineteenth century. There appears to be ample opportunity for postgraduate research in engineering and the natural sciences, and our data suggest that in engineering a higher proportion of graduates go on to complete a doctor's degree than is the case in Britain.

Below the universities there is the range of vocational and technical courses offered by the *Fachhochschulen* leading to the *graduiert* qualifications — in engineering to the Ing. Grad., of course. At the same time there now exist some *Gesamthochschulen* bridging this university *Fachhochschule* gap, and seeking to offer the advantages of both institutions (see Chapter 2).

Then below the *Fachhochschulen* are a complex of other institutions of further vocational and technical education (again see Chapter 2). These variously serve to provide postsecondary-school education, introductory vocational courses, courses for apprentices, for would-be foremen, for technicians, for works-study personnel, and courses leading to higher level institutions within the system, including the *Fachhochschulen*.

Similarly in this connection it should be underlined that apprenticeship is taken very seriously in Germany, and there are banking and commercial apprenticeships as well as craft and technical–industrial apprenticeships. This claim may be supported in several ways. First, there are a large number of occupations served by an apprenticeship scheme in Germany, over 460. Second, the apprenticeship system is administered and monitored by the *Industrie-und Handelskammer* on a local basis, but a final written examination is always required as well as an actual demonstration of craft skill. Third, being a *Facharbeiter* (skilled worker) is a legally protected status and is co-terminous with the successful completion of an apprenticeship. Fourth, admission standards for apprenticeship training have risen, and it is now common to hear German personnel officers saying that they are looking for people from selective secondary schools with the German equivalent of 'O' level. In particular the apprenticeships offered by the 'big three' banks are especially sought-after and are a perfectly acceptable career choice for someone with the German equivalent of 'A' level. Fifth, in the tradition of the American founder of scientific management Frederick Taylor, it is possible to do double apprenticeships (in different skills and subjects) and not uncommon actually to do so. Sixth, doing an apprenticeship and becoming a *Facharbeiter* is usually a prerequisite for subsequently enrolling for the foreman's certificate (*Meisterbrief*) course, and eventually becoming a foreman. To round off this sketch of apprenticeships in Germany it might be added that the Germans have also taken up the idea of training for semi-skilled jobs, designating some of these as *Anlernberufe* (semi-apprenticeships) and evolving a systematic training scheme for them.[3]

The thrust of this part of the argument and these further references to the German education system is to indicate that from apprenticeship to Ph.D. the German system has accommodated itself to the needs of industry. Or to put it more modestly, vocational education, at several levels, is a strength of the German system. If industry did not have high standing in Germany this would not be so.

We would like to conclude this chapter in terms of a precise comparison which may serve to illuminate the role of industry in German life. By way of introduction, and after the former discussion of history and education, we will start with something tangible — money. In March 1976 *Mechanical Engineering News* published the results of its salary survey of the members of the Institution of Mechanical Engineers.[4] One of the tables in this source is a breakdown of salary by type of employment, and it is very instructive in the present context (see Table 8.1).

The actual salary figures, of course, are not up to date, but

TABLE 8.1

Remuneration of IME Fellows and Members,
all ages in median annual salaries by
class of employment 1976

	£
University	6,970
Hospital Board	6,830
Armed Forces	6,750
{ Central Government	6,350
{ UKAEA and associated companies	6,350
All public sector employment	6,230
Nationalised industry or public corporation	6,130
Self-employed	6,000
Any other employer	5,980
Local Authority	5,874
Industrial or commercial company or firm	5,650
Consultancy Practice	5,300

the absolute amounts are not the important thing. What is important is the ordering in which every kind of public sector employment is better remunerated than work in industry. Our German salary data, dating from the same period, can be cross-tabulated with type of employment in a similar way. We did this and found that engineers working in industry were paid more, on average, than those in all other types of employment, and this was true for both Dipl. Ings. and Ing. Grads. Exactly the same emerges from the survey conducted by Eugon Kogon, discussed in the previous chapter in connection with the socio-political attitudes of engineers. Kogon similarly reports higher average remuneration for engineers in industry for all qualification levels, and all age-grades. And his sample was over 25,000.

Obviously there is nothing absolute about these two sets of salary relativities, but they are quite consistent with the opening proposition of this chapter, that the status of industry is higher in Germany. This proposition finds some support, too, from a more sustained comparison of the two types of engineer in Germany, those working in industry, and those working in the public sector, and it is to this that we now turn.

This exercise has to some extent been anticipated in that several of these industry–public sector contrasts and divergences have been mentioned in earlier chapters as they were relevant to particular themes. It has been made clear, for instance, that the three sub-groups — those working in industry, those with good degrees, and the successful — overlap to some extent, and share some attributes. There are, however, more ways in which the industry and public sector groups may be differentiated.[5]

The industry group, in declaring the reasons which originally attracted them to a career in engineering, mentioned less often than their public sector colleagues an interest in natural science. But they mentioned more frequently being influenced in their choice of engineering by fathers who were engineers. This is interesting especially in conjunction with the finding mentioned in the last chapter, namely that the children of engineers working in industry are themselves more likely to aspire to become engineers than are the children of the public sector group. The industry group were more likely to have done an apprenticeship than were the public sector group; this is not a truism given the role of the apprenticeship in the German vocational system — 71 per cent of the whole sample, in fact, had completed apprenticeships. The industry group were also less likely to indulge in retrospective criticism of the apprenticeship. With regard to their college courses, however, a higher proportion of the industry group said they would have liked both business economics and foreign languages included in their course. The industry group were also more enthusiastic about the *Praktikum* (the compulsory training period in industry) than engineers working in the public sector (though the majority of both groups were positive on the subject of the *Praktikum*). In fact the industry group were more likely to view their college education as 'very good' — a propensity they shared with the successful and those with good degrees.

As was the case with those engineers designated as successful (see the discussion in Chapter 5) the industry group had more authority. They emerged, that is, as more likely to be able to take decisions in a range of fifteen functional areas — production, design, and so on — though this in part denotes the difference between manufacturing (industry) and the provision of services (public sector). Those in industry were also more likely to have authority to appoint other qualified engineers, and among those engineers with responsibility for immediate subordinates, the number of subordinates was higher for those in industry.

There were also some interesting differences in attitude. Those at present employed in industry were more likely to have considered an eventual career in management at the time they took the decision to become engineers. The industry group were also more likely to say that they would respect more an engineer who became a *Vorstand* member[6] than one who made a significant contribution to research. These pro-industry-management attitudes of those at present employed in industry, however, were not to the exclusion of engineering itself. When asked whether, if starting out in life again, they would still choose engineering, a higher proportion of those in industry said they would. And in response to the further question, about whether they would choose mechanical engineering again, a higher proportion of the industry group again affirmed that they would. The engineering-based group also showed greater satisfaction with their present jobs, nearly 40 per cent of them describing themselves as 'satisfied in every-way'. The industry group showed a correspondingly higher level of satisfaction with the general status of the engineer in German society. It is also of interest to note that those in industry were more prone to believe that the attitudes of their non-technical colleagues towards engineers were favourable. In all these questions of attitude and preference it will be noted that the industry group have the same disposition as the successful (see Chapter 5).

Again, like the successful, the industry group expressed a higher predilection for jobs with good promotion prospects rather than security (though the majority preference was for jobs offering a modicum of both). In response to a question about the relative importance of formal knowledge and experience the industry group were a little more likely to emphasize experience. They attached more importance too to the role of originality in the solution of engineering problems. They were more likely to be able to speak foreign languages and more likely to have patented inventions to their credit. And the industry group were more likely than their colleagues in the public sector to claim that they had made some contribution at work going 'beyond the call of duty', and were able to offer examples. They also worked longer hours, actual not prescribed. Correspondingly the industry group spent less time on gardening, jobs around the house, and working on their cars — like the successful.

Finally, as was said in the earlier discussion of the engineer's non-work life, their children are over-represented at the specialist economics and technical grammar schools, and more likely than the sample average to want to become engineers themselves.

We have worked through this set of contrasts in some detail for two reasons. The first is that the configuration of elements in the make up of the industry-based German engineer is interesting. It is comprised of greater business, responsibility, and sacrifice, and higher morale in the sense of greater commitment to engineering and the job, more positive evaluation of earlier training, and so on. The picture is also suffused with a strong element of practicality: the apprenticeship, foreign languages, the evaluation of the *Praktikum,* and the record on patents. They are richer, and on average younger, and they have more in common with the successful and those with good degrees than with the other sub-groups. Finally, they are exerting a stronger influence over their children in favour of engineering than are their colleagues in the public sector.

The second reason for having worked through the details of this contrasted portrait is more speculative. It is to suggest that this is the kind of engineer that will be attracted in a society in which industry does enjoy a high status.

Chapter 9

The Status of Engineers

In January 1980 the Report of the Committee of Inquiry into the Engineering Profession, chaired by Sir Montague Finniston, was published.[1] Two quotations from this source will serve to set the scene for the theme to be explored in this chapter.

In one of the summary sections the Report declares:

It is clear that, unlike their counterparts in other industrial countries, engineers in Britain lack the special social standing which attracts young people to aspire to an engineering career, and that they are ill-served by a generic title which in Britain is not specifically associated with and reserved to a highly educated and vital professional group. Engineering is further regarded misleadingly as a branch of science, rather than as a culture and activity in its own right.

In a discussion of the findings of members of the Committee on their visits to several foreign countries the following observation is offered on West Germany:

Although German engineers are found in increasing numbers outside engineering functions, this is not because engineering offers less attractive career prospects. On the contrary, we formed the impression that engineering is regarded as an attractive career both in status and in more tangible terms. The senior managers of German industry tend to emerge through the engineering function and about 60% of board members of German companies have engineering backgrounds.

These two quotations indicate a significant contrast; the thesis of the present chapter is that engineering and its practitioners enjoy high standing in West Germany, and a higher standing than in Britain. This thesis is a logical companion of that concerning the higher standing of industry in West Germany argued in the previous chapter, and it is also intended here to make clear some of the connections between these two contentions.

More modestly, however, we will begin with two qualifications to the main argument concerning the higher status of engineering. The first is that German readers may feel we have overstated the case. It is noticeable indeed that older engineers in Germany are inclined to take a 'things ain't what they use to be' line on the standing of the profession, and it is possible that they are right — it is, of course, a difficult hypothesis to investigate. It has also sometimes been noted that the standing of engineers, as measured by their relative position in surveys where respondents are requested to put a number of jobs in order of status, is subject to short term fluctuations.[2] While such modifications of the main contention might be conceded, the counter-argument we would offer to any doubting Germans is that this thesis is being developed from a British standpoint, and by British standards German engineers are both materially and culturally 'well-off'.

The second qualification concerns the rules of evidence and logical inference. It is difficult to prove, or for that matter disprove, global propositions about the social standing of occupational groups. This is partly because of the pure complexity of such issues: any occupation has many features and characteristics, inherently and in relation to other occupations, all or any of which may serve to raise its relative standing in a given society, and it is partly because the sort of arguments one will use to justify a claim about an occupation's status often turn out to be reversible, and thus open to objection from the logical purist. Things like the German engineer's access to management posts, for instance, cited earlier in the second quotation from the Finniston Report, may be said to cause or contribute to the status of the engineering profession, or to derive from it, be determined by it. It is the same with many of the phenomena we wish to discuss here. To put it another way, the logical problem means we are discussing aspects of a phenomenon rather than finite causes.

In the previous chapter it was argued that the German system of further and vocational education is markedly adapted to the needs of industry, and that this reflected the central position and high status of industry in German society. To open up a variation on this thesis we would like to suggest that some features of the German education system, and the place of technical education therein, may be cited as evidence of the standing of engineering.

To take a straightforward comparative point, the proportions of students studying various subjects at university in Britain and Germany are different — and in Germany engineering is one of the favoured subjects. The British strengths, in the sense of subjects where the numbers of students form a higher proportion of the total undergraduate population than is the case in Germany, are the traditional humanities and natural sciences. On the German side the corresponding strengths in this sense are a range of subjects having a vocational significance — economics, business economics, law, and engineering. The proportional difference where engineering is concerned is fairly modest but in Germany's favour. The most striking difference in the comparative league concerns law, which is almost certainly a manifestation of the vocational emphasis in higher education in Germany. The study of law, it should be added, does not just lead to the practice of law in Germany; it is the subject for entry to the higher civil service, and a subject for industrial managers.

Again on the quantitative approach to the role of engineers in the system it is worth looking at the British manpower planners' QSE (qualified scientists and engineers) construct. In both countries it is the case that engineers rather than scientists constitute the majority of this group but the engineers' majority in the German QSE group is about 20 per cent larger than in Britain. This QSE data is consistent with one of the findings of the Finniston Committee, in turn gleaned from Unesco Yearbooks, to the effect that German engineering graduates constitute a higher proportion of the relevant age group (see Table 9.1).

It is interesting to note with regard to this table that Britain is not exceptional in either direction; the two 'odd men out' are in fact the 'economic terrors' respectively of Europe and the Far East.

There is of course a qualitative aspect to this as well. For some years it has been a common belief in Britain that the typical engineering undergraduate has a sub-average 'A'-level attainment. Unfortunately it is not just a question of 'common belief' — there is some evidence for this proposition. One of

TABLE 9.1

Engineering graduates as a proportion
of relevant age group

Country	date of datum	proportion
UK	1978	1.7
SWEDEN	1977	1.6
FRANCE	1977	1.3
USA	1978	1.6
GERMANY	1977	2.3
JAPAN	1978	4.2

the present authors tried, a few years ago, computing what proportion of students reading various subjects had a very good 'A'-level attainment (3 subjects, all A and B grades) using the statistical supplements to the UCCA annual reports. The qualitative picture which emerges from this exercise is similar to the quantitative: maths and physics, classics and other humanities subjects, and medicine, are in the top quartile; civil engineering, mechanical engineering, electrical engineering, and business studies are in the bottom quartile. A report to the Co-ordinating Group of the British Association in 1977 also made it clear that the average 'A'-level attainment of British engineering undergraduates was below the average for the undergraduate body as a whole.[3] It should be added that it is probable, but not certain, that the situation in Britain in this respect is changing for the better; certainly the testimonies of some individual professors of engineering suggest that this is the case.

It would be useful to run this same check, on the average *Abitur* (German equivalent of 'A' level) attainment, that is, of undergraduates reading various subjects including engineering. Unfortunately this sort of datum is next to unobtainable in Germany. The ZVS (*Zentralstelle für die Vergabe von Studienplätzen*, the German equivalent of UCCA) was founded somewhat later than UCCA in Britain and it certainly does not publish statistics which would enable calculations similar to those we made on the basis of UCCA data. Reluctantly one is forced to the conclusion that the non-availability of these data is an indicator that Germans think less about relative status, even if certain groups in German society, including qualified engineers and industrial managers, have more of it. A further complication concerning the ZVS is that it is a federal institution but education is a *Land* (state) matter. Thus the ZVS is in the position of a data-holding and administrative trustee for the *Länder* (states) which again circumscribes its position.

Thus it is not possible to match the British data exactly, but we have managed two checks on *Abitur* grades from subject to subject. The first of these derives from the personnel records of a multinational German firm. This company collected data on the *Abitur* grades of its graduate applicants, and thus is able to state the *Abitur* grade averages for (its applicant) graduates in mechanical engineering, chemistry, law, economics, and business economics.[4] The range of subjects is limited but it is an interesting group for our purposes. In the sample of applicants for posts with this company the chemistry graduates had the highest average *Abitur* attainment, while the four other graduate subject groups were all about the same and putatively in the middle of the German national spectrum. Now in Britain, by contrast, the law undergraduates would have the highest average 'A'-level grades and be in the top third of the subject league in this respect. The chemistry undergraduates would be next and roughly in the middle of the league, while the students of mechanical engineering, economics, and business studies (our nearest equivalent to the German business economics graduates in the company sample) would all have an average 'A'-level attainment, placing them in the bottom third of the league. We could put this another way and say that if our mechanical engineering students come, in the future, to have average 'A'-level grades equal to those of students of law, the engineering faculty admissions tutors will be pleased — and the country as a whole ought to be pleased.

The second factual sidelight on the relationship between university subjects and *Abitur* grades relates to the city of Brunswick. With the help of a German friend we were able to check the average *Abitur* grades of those going on to university to read engineering against the average *Abitur* grades for all students who passed the *Abitur* in that city (population: 240,000-plus) in one year in the late 1970s. It emerged that the would-be engineering students had an average *Abitur* attainment above the average for the city's students as a whole.

It is perhaps also relevant to refer in this context to hearsay evidence. That the typical engineering undergraduate in Britain is less likely to have passed three 'A'-level subjects, and less likely to have good grades, than, say, the typical physics undergraduate is, unfortunately, something which has been common knowledge for some time, even if the situation is improving. There is no analogous piece of common knowledge in Germany, and if one asks those who are in a position to know, the general view is that engineering students would have an average *Abitur* attainment or be somewhat better than the all-subjects-combined average.

The foregoing discussion does not indicate that we are obsessed with 'A'-level grades and their German equivalent, nor that we are unaware of the literature which reports on investigations of the correlation between 'A'-level grades and later degree performance and finds that 'A'-level grades are often not a good prediction. This point of emphasis is rather that these grade questions are particularly apposite in discussions of status. After all, if we take the English half of the equation, sixth formers who get 3 A's at 'A'-level are the ones considered the cleverest at that stage — by their teachers, their class mates, by university admissions tutors, and arguably by themselves. What they choose to study is of indicative interest, and, in the light of our early 1970s' data at any rate, they put classics top, business economics bottom, and the engineering subjects not much higher. This would be inconceivable in Germany.

It is also reasonable to argue that the German engineer's high absolute and relative remuneration is a reflection of his standing in German society. Some data on German engineering salaries have already been given (Chapter 4). It might be added that German engineers are absolutely better paid than their British colleagues even after allowing for the higher cost of living in Germany (some precise comparative data on this point are offered in Chapter 10). It may be felt that this is not a very telling piece of information since the German GNP is getting on for double the size of the British GNP although the population difference is not large. Using the measure of GNP *per capita,* West Germany is the richest member of the EEC, and for that matter one of the richest countries in the world. Thus a great many occupational groups in Germany are better paid than their British equivalents.

It is, however, possible to go further: there is some evidence which suggests that German engineers are better paid relatively than are their British colleagues — compared, that is, to other middle-class occupations. One study, for instance, suggests that the German engineer is more highly remunerated on average than the university professor and the senior civil servant, whereas in Britain both the professor and the civil servant, earn more on average than the engineer.[5] We have also been given some salary data by the *Bundesanstalt für Arbeit* (federal labour office) indicating that the average remuneration for German graduates in mechanical engineering is superior to that for university graduates (all subjects) in general.

It is a *leitmotiv* of this chapter that the (high) status of both industry and engineering in Germany are interconnected. The most obvious connection is the access, one might almost say dominance, of engineers to management posts. Every survey of the background and qualifications of German managers shows engineers to be the most numerous (subject) group.[6] Engineers are well represented on the executive boards (*Vorstand* or *Geschäftsführung*) of German companies, and the common British phenomenon where there is only one director qualified in engineering but several with sales or finance qualifications and experience, is unusual in Germany in our experience. Finally qualified engineers in German companies completely dominate the technical functions — R and D, Design, Engineering, Production Control, Production, Quality Control, Testing, and so on, and 'overspill' into the non-technical functions. It is interesting to note that in Germany the reverse does not seem to happen: people with non-technical qualifications, that is to say, do not 'overspill' into the technical functions. Apart from the strengthening of the engineer's status, it is worth adding, this strong representation of engineers in the management of German companies also means that there is an omnipresent lobby for design, quality, and technical excellence.

It might at this stage be helpful to move to a more general

cultural consideration, concerning the way in which branches of knowledge are perceived and related. The simplest and most popular classification system, not just in Britain but in the English speaking world generally, is the idea of the two cultures, with the arts or humanities subjects on one side of a divide and the natural sciences on the other. This Anglo-Saxon thought pattern, which counterposes Arts and Science, is not really conducive to the dignifying of engineering. This is partly because engineering does not fit unequivocally under either of the 'two cultural' headings, and partly because by labelling engineering 'applied science', the usual way out of the impasse, one is assigning engineering to a subordinate and dependent status.

It may be argued that engineering should not be viewed or labelled as 'applied science' for several reasons. A basic consideration here is that the output of science is knowledge, whereas the output of engineering is three-dimensional artefacts. Also, the relationship between science and engineering is variable, and certainly not isomorphic. Advances in engineering, that is to say, are sometimes dependent on, and subsequent to, prior advances in science, and sometimes the advance in engineering is quite independent. Another difference is that engineering practice is much more subject to environmental and economic constraints; it is, in other words, more concerned with workable solutions than with ideal solutions.

German thinking on the perception and classification of branches of knowledge offers an interesting contrast. The two-cultures distinction does not exist in Germany, and the idea is indeed difficult to formulate in German. It is possible to translate 'applied science' into German, but the result of this endeavour is culturally meaningless. The Germans have a three-fold classification scheme: this means not only that they have a 'third culture' but also that they draw the boundaries in different places.

The German term *Wissenschaft* covers all formal knowledge subjects, whether arts, science, or social science in our terms. This explains the rather casual use of the word 'scientific' by Germans when speaking English: in their view it can be applied as readily to historical scholarship as to nuclear physics. In the

German scheme *Kunst* denotes art — not 'the arts' in the Anglo-Saxon sense, but the 'products' of the arts. The criterion for inclusion is aesthetic not critical–intellectual. And the 'third culture' in the German scheme of things is *Technik*. *Technik* is for the Germans an independent domain, embracing knowledge and skills relevant to manufacturing. Thus it is an autonomous cultural rubric tending to dignify engineering, and certainly serving to differentiate it from natural science.

Furthermore the word *Technik* is very homely in German; its use is certainly not the preserve of philosophers of science or epistemological theorists. People in Germany speak of liking *Technik* or working in *Technik*. In industry it is a generic term tending to cover all those associated with manufacture from the machine operator to the technical director. Bodies of knowledge and special skills relating to particular manufacturing processes are usually denoted in German by compound nouns ending in *Technik*. The special character of German democracy, indeed, means that anyone is allowed to originate *Technik* compounds in German.

We have argued here that the German idea of *Technik* tends to enhance engineering, and certainly to save engineering work from both misapprehension and demotion under the label of 'applied science'. A similar Anglo-German contrast may be developed with regard to engineering and the idea of the profession. This theme was touched on in the Introduction where it was suggested *vis-à-vis* British engineers that their desire to be regarded as a profession is readily comprehensible, since it would be status-enhancing, but is in fact controversial. If one takes the idea of 'the profession' precisely, and argues from the characteristics of say law and medicine, which certainly do rank as 'professions' in the English scheme of things, then certain factors may be adduced in favour of the engineers' aspirations. These include the facts of high education, lengthy training, the application of specialist knowledge and skills, responsibility, and social utility — a powerful battery of arguments in favour of engineering. On the other hand, and continuing to use the traditional professions as a model, it has to be admitted that self-regulation by a professional peer group is weaker with engineering. Nor are most engineers 'independent professionals' after the manner of lawyers; engineers tend to work for large organizations, typically manufacturing companies. This means that (most) engineers cannot be guided by criteria of technical excellence alone, but are subject to economic and managerial constraints. Again, and especially in the English-speaking context, becoming an engineer is not necessarily a terminal career state: many engineers aim and do become general managers, and are pleased to be known as managers rather than engineers. This again is incompatible with the purist concept of the profession.

Thus for the British engineer the issue is contentious. But if he did have an incontrovertible claim to being a professional this would raise the general standing of engineering in British eyes, where the established professions are all high prestige occupations. The arguments for and against the British engineer's claim apply equally to the German engineer. The key difference is that the Germans do not really have a concept of 'the professions' so German engineers cannot lose any increment of status by not having an undisputable claim. The German word for 'profession' is *Beruf,* but this is also the basic and general word for job or occupation, so that the Germans do not have a linguistic mechanism for singling out certain (professional) jobs and endowing them with an added element of status. To complete the German picture it should be said that the word *Beruf* does have a middle-class ring to it and a secondary connotation of respectability. On the other hand the particular job does not have to be professional in the Anglo-Saxon sense to qualify under the respectability clause: a sales manager is as good as a rural dean, a design engineer as good as a solicitor.

There is also a broadly historical dimension to the higher standing of engineering in Germany. The central consideration here is that in Germany the state has been much more closely identified with engineering than it has been in Britain, and particularly with the development of technical education. Germans are not conscious of this any more, and 'the state' has in any case lost its aura. The German engineer, however, continues to enjoy the status which is a legacy of this development. Several factors relating principally to the education system derive from this traditional state initiative in Germany, and there is here some contrast with the results of the initiative exercised in Britain by the various engineering institutions.

First, in Germany the institutions of higher engineering education, the *Technische Hochschule* and the *Ingenieurschule*, which preceded the present *Fachhochschule* (see Chapter 2), were all established by the state. Furthermore, many of the *Technische Hochschulen* were established in the nineteenth century and had parity of esteem with the traditional universities. Engineering courses at both these levels are full-time, and include organized 'practicals' in industry. Such 'practicals' are often part of the corresponding British courses but not invariably so: in Germany the state initiative has made this a uniform requirement. Again, as was made clear earlier (Chapter 2) the German courses are not only full-time but longer than the corresponding courses in Britain: this fact indicates the importance the German state has attached to engineering education.

Another by-product of the state design and control of engineering education in Germany is that the qualification system has been very clear cut and straightforward. Above the technician level there is the *Fachhochschule* trained Ing. Grad. and *Technische Hochschule* trained Dipl. Ing., and no further complications. This simple and clear-cut system has probably contributed in a small way to the German engineer's standing. It has meant that not only employers but also the general public has readily understood what the two grades connote and how they equate with non-engineering qualifications. Thus, even if they had been disposed to such a tendency, it has been that much more difficult for Germans to misconstrue these engineering qualifications, demote them, or confuse engineer with mechanic.

This pristine picture given in the previous paragraph should be modified a little in the light of the establishment in the last ten years of a few *Gesamthochschulen* (see Chapter 2) with their H and Y model courses combining in part the traditional Dipl. Ing. and Ing. Grad. streams. Even with this qualification, however, a comparison with Britain remains. In the recent past we have had three equivalents to the German Ing. Grad. — the Technician's Certificate (Parts I, II, and III), the Higher National Certificate, and the Higher National Diploma. At university level we have the distinction between traditional and sandwich courses, and between ordinary university degrees and CNAA degrees. To this should be added the fact that the British engineering institutions have also offered qualifications and membership grades having a qualificational significance. The system has changed with time, and so have the titles. The man on the Düsseldorf tram knows the difference between a Dipl. Ing. and an Ing. Grad., but it is doubtful whether the man on the Clapham bus could distinguish between an M.I. (Mech.) E. and a C. Eng., or between an HND and an HNC.

We have argued that the initiative of the state, and the esteem in which the German state used to be held, have had both specific and general effects on the standing of engineering in Germany. Another cultural factor which tends to underpin the position of the engineer is the German predilection for specialization.

One easily identifiable aspect of this German penchant for specialization is the much closer connection between subject studied at university and later occupation. The young German on graduation day has much less choice among possible career and job openings than his British colleague. The German, to a much greater extent, chose his subsequent career when he decided on his course of undergraduate study. The difference may be highlighted by offering for a moment a German view of Britain: Germans find the idea of a classics graduate entering the higher civil service or a history graduate becoming a management trainee quaintly amusing. The German view of the proper relationship between university and occupation is reflected in the different proportions of students studying the various subjects which were referred to earlier in this chapter. Another aspect of this German preference for specialism is the strength of vocational education in Germany, a theme which has recurred in this book.

It is also fair to say that there is an intangible but real aspect to this German feeling for specialism. It is noticeable that Germans think in specialist terms on a wider occupational front than we do, and they continue this specialist thinking further up their organizational hierarchies. One of the present authors once had occasion to visit a range of social science research institutes in West Germany, and asked among other things what were the further job possibilities for younger research staff if they left the institute. The answers gave some evidence of this specialist thinking. They were not on the lines of: research fits one for this but does not fit one for that, but tended to be much more specific *and* to be differentiated according to the kind of research being conducted. Thus one would receive quite different answers at an education research institute from those at a economics institute.

The point of these remarks is not so much to favour a specialized rather than a general training as to note the emphasis in Germany, together with the fact that the German tendency to specialize favours the engineer. This is for several reasons. First, and most obviously, the engineer *is* a specialist. Second, his specialism *has* practical application, in a way that of say the classics graduate in higher administration does not. And third, this specialism gives the German engineer an *a priori* right to enter industrial management, in a society in which industry also enjoys high standing (see Chapter 8).

These remarks on the German preference for specialism lead us to consider another aspect of German society. It is more materialistic than Britain, or at any rate more frankly materialistic. This feature is not as obvious now as it was in the 1950s or 1960s but it is still observable. Germans care about personal wealth and possessions more widely and conceivably more deeply than the British. Germans also identify much more strongly with the over-all economic performance of their country. These traits are very easy to understand. The Germans, in the immediate post-war years, experienced poverty and deprivation on a scale unknown in other western countries. Their great achievement in thirty-five years was an economic achievement. And this was a sustained performance: oil crisis and world recessions came and went, but Germany went on apparently for ever with its tiny inflation rate and enviable trade surplus. The economic success of West Germany has another simple importance for the citizens. It is their personal victory over the East, it shows that West is best, and that the Berlin Wall was not needed to keep the westerners in their place.

Side by side with this putatively greater materialism exists a more marked tendency towards meritocracy in West

Germany. Propositions of this kind are, of course, difficult to prove, and it is admitted that they are somewhat impressionistic. It is certainly possible, however, to give reasons for this view and illustrations of it.

This German enthusiasm for meritocracy is clearly discernible in a negative way: there are fewer ascriptive, or semi-ascriptive, bases of prestige in Germany. There are, for instance, no public schools in the English sense, no Oxbridge, and no equivalent of the London clubs (Bonn, after all, is only an ex-market town). There is also no monarchy; the standing of service officers is low since the Second World War, and there is no indigenous word for 'gentleman'. No prestige attaches to residence in any particular part of the country: there are no home counties, no golden circle, no Washbos or Chipitts,[7] and no (one) gin-and-jaguar belt. Indeed in terms of real geography, as well as the political and administrative orthodoxy, West Germany is the ultimate in decentralization. Berlin is the ex-capital and largest town, but it is isolated and has a declining population. Bonn, the new capital, is not much bigger than Southampton, and even lacked an airport until the RAF gave up their base at Wahn. The largest 'mainland' town is Hamburg in the North, closely followed by Munich in the South (sometimes described as the 'unofficial capital' because of its scenic attractions and expanding population). The centre of gravity of both industry and population is *Nordrhein West-falen,* and the commercial capital is Frankfurt. The towns of Berlin, Hamburg, Hanover, Düsseldorf, Frankfurt, Stuttgart, and Munich are in every sense regional capitals: there is no German equivalent of Paris or London. Germans, like Americans, claim that their accents are regional rather than class-based.

To put it another way, there are several senses in which German society is rather undifferentiated. What, then, may distinguish the individual? Our impression is that it is wealth, power, and achievement, and these to a higher degree than in Britain.

There are several positive manifestations of this. The preference for specialism is probably one: specialist qualifications render their possessors rationally and morally fitted for particular positions in the occupational structure. The fact that the word ambition is on the whole used positively is another. The prevalence of the word *Leistung* (performance and achievement) and its numerous compounds and derivatives including *Leistungsgesellschaft* (achievement society or meritocracy) is another.

The purpose of these observations on German materialism and meritocracy is again to suggest that they tend to favour the engineer and underpin his standing. These tendencies will favour the engineer because he will be correspondingly valued for what he knows, what he can do, and what he can contribute to national wealth — and not for whom he knows, where he comes from, his cultural image or social deportment.

In the previous chapter it was argued that industry has high status in Germany. Obviously the German engineer benefits from this association, indeed the status of industry and the status of engineering are mutually reinforcing. A particular feature of this industry–engineering alliance is the fact that there are considerable institutionalized connections between industry and those institutions where engineers are trained. There is the universal institution of the *Praktikum* (see Chapter 2) which takes all engineering undergraduates into industry for at least six months. Another (frequent) tie derives from the various projects which are mandatory on both the Dipl. Ing. and Ing. Grad. courses. These projects may be either theoretical in nature or empirical. They are more often empirical, and often constitute investigations of problems of direct interest to industry. Sometimes the experimental work is done in industry itself, rather than at college, and these are the sorts of contact that often lead to first placements.

One of the present authors once had the opportunity to visit a number of *Fachhochschulen, Gesamthochschulen,* and *Technische Hochschulen* (all the types of educational establishments at which higher engineering training is offered, see Chapter 2) and seized the opportunity to ask about contacts with industry. All interlocutors in fact claimed substantial contacts: this may not be thought a very devastating finding — one would hardly expect heads of engineering colleges to say they had no contact with industry (though we have on occasion heard this claimed with some pride in Britain). If, however, one asked as a follow-up question for examples of such contacts the normal response was an impressive litany of information exchange, personal contacts, student placements, exchange of personnel, research contracts, joint problem definition, and so on.

Finally in this context of industry–education links there are considerable overlaps of personnel. The appointment of engineers based in industry as part-time university lecturers is probably more widespread, and of longer standing, than in Britain. It is also true that the typical engineering professor in Germany has a fair amount of industrial experience. The typical route to a chair in engineering in Germany would run like this. Undergraduate study, followed by full time research for the Dr. Ing. (Ph.D. in engineering). At this point one would normally leave the university and go into industry. Not just for an experience-gathering eighteen months but for a substantial period of about six to twelve years. During this period one would continue to research and publish, and re-entry to the university system would eventually take place at full professor level. This is not the invariable pattern, but it is common — and comprehends all the engineering professors we have actually met. This industry–education contact is even more rigorously preserved at the *Fachhochschule* level where the Ing. Grad. is trained. In order to be a lecturer at a *Fachhochschule* one must be a university graduate in engineering, a Dipl. Ing., that is, and have a minimum of five years experience in industry. These requirements are embodied in the state law.

Finally there are perhaps two other factors relevant to the standing of the engineer in Germany, and they both derive from the War and its effects. The first is that the engineer was very much the 'man of the hour' in the immediate post-war period. The purely physical destruction of Germany in 1945 was enormous in extent and the call for (literal) reconstruction very urgent. Eventually the task of reconstruction in a literal sense gave place to that of refashioning post-war Germany as a competent member of the world economic community. In both enterprises the German engineer contributed handsomely; his contribution particularly to the first phase turned him into something of a folk-hero.

The second and final point again concerns Germany's position in 1945, in what Germans call '*die Stunde O*' (the hour nought, the absolute zero). Germany in 1945 was perhaps like no other country has ever been. Militarily defeated, morally condemned, politically emasculated — and soon to be starving. Germans in those days could not hope to play a part on the world's stage; they could scarcely aspire to political independence, still less to a military recrudescence. Even German culture, with its sometime *Herrenvolk* connotations, stood condemned. As the post-war years went by, however, the Allies softened in one respect. They were prepared to see Germany recover economically. At first it was a case of German recovery reducing the obligation of the Western Allies to feed them; then, especially as western relations with the USSR deteriorated, the Allies came to favour an economically strong, stable, and viable West Germany. In short the only form of achievement open to the Germans after the Second World War was economic: at first grudgingly permitted, then roundly extolled. The Germans were equal to this uniquely devised opportunity. At first it was seized in a spirit of self-help; later it became a vehicle of international self-respect; finally a surrogate religion. And it cemented as no other historical contingency could, the alliance of industry and the engineer.

Chapter 10

Learning from Germany?

In the Sheffield City Art Gallery there is a painting by William Roberts, dating from 1934. It depicts some friends in an artist's studio regarding the picture painted by one of the group; another member is clearly expostulating. The title is in the form of a direct speech caption:

'No! No! Roger, Cézanne did not use it.'

Which is a way of saying that in the use of models, including international ones, slavish emulation is not recommended. German engineering standards and achievements are well respected, and the country's economic performance is impressive, but these are not reasons to introduce the Umlaut (¨) into English or begin driving on the right hand side of the road. In other words, the interest is in highlighting the best, not copying the lot.

There may also be a case for establishing with more clarity the 'learning gap'. The question of our learning from Germany, that is, can only be raised if there are significant differences between the two countries. One of the themes of this book has been that German engineers, and their situation in their own country, are sufficiently removed from British experience to be worthy of discussion. Many of these differences, both of detail and ethos, have been mentioned, but a summary statement may be helpful at this point especially if accompanied by contrasting data for Britain.

The last point, providing some British data to set beside the facts about Germany, including some derived from our survey, is eminently possible but is subject to a qualification. As was explained in the Preface, the survey of German engineers was modelled on that already conducted in Britain by one of the present authors together with an American colleague.[1] The two surveys, in Britain and Germany, are really very close. Essentially the same questionnaire was used in both studies, with some additions and modifications in the German survey;

the samples were of similar size, and of similar composition with regard to the age structure, qualification mix, and occupational deployment. The qualification is that there is a substantial lapse of time between the two studies, the British field work dating from the early 1960s and the German from the mid-1970s. It is important to make this qualification clear because in a few of the contrasts, presented below, which are in the main in Germany's favour, a later and synchronic survey in Britian would probably have reduced the gap. To give a straightforward practical example a higher proportion of German engineering undergraduates did doctors degrees after their first degree, but the proportion probably went up in Britain as well in the time between the two surveys. It should be added that the qualification which is made clear here does not by any means apply to all the contrasts which follow: many are outwith the two surveys, pertain to the training systems in the two countries, or are cases where it was possible to match data in time. This last applies to a contrast on remuneration, the first item in our summary list:

ENGINEERS IN BRITAIN AND GERMANY:
SUMMARY DIFFERENCES

1. The salary survey conducted by the British Institution of Mechanical Engineers in January 1976 showed a gross median income, all age groups, membership grades, and qualificational levels combined, of £5,510 p.a.[2] Our own survey in Germany from the Spring of 1976 showed a median gross income of £12,119 p.a. (converting at 4 DM: £1). The cost of living was somewhat higher in Germany at the time but by no means so much higher as to cancel out this vast difference in salary. In fact according to the United Nations cost of living index for mid-year 1976 it was 36.4 per cent higher in Germany.

2. There is also some reason to believe that the German engineer is better paid in relation to other middle-class occupational groups in his own country, than is the engineer in Britain; in this connection a particular study was cited in Chapter 9.[3]

3. Using data from the same year (1976) it was further demonstrated (Chapter 8) that for the engineer in

Britain all species of public-sector employment appear to be better remunerated than industry, whereas in Germany the average remuneration of engineers in industry was appreciably higher than that of their colleagues in the public sector.

4. This last fact of higher remuneration in industry in Germany, together with numerous other contrasts concerning German engineers working in industry and the public sector, was adduced (Chapter 8) to support our claim that industry enjoys high standing in Germany. By common consent the standing of industry has not been high in Britain.

5. The training period for engineers in Germany is somewhat longer than for their colleagues in Britain; this is especially true of the Dipl. Ing. where the official minimum course length is four years, and the actual time taken to graduate is nearly six years. The German university course also includes a mandatory six months in industry (the *Praktikum* described in Chapter 2).

6. If we compare the engineers in our German survey with those in the earlier British study it would appear that more of the former come from middle-class homes (see Table 10.1).

TABLE 10.1

Father's occupation classed as	UK graduates	BRD per cent	UK non-graduates	BRD per cent
professional or executive	40	58	20	30
middle white-collar	36	22	34	26
manual	22	18	43	38

This data on class background is not advanced as 'a point for the other side', but it is interesting in the light of the argument developed in the last chapter that the engineer enjoys higher standing in German society. The more middle-class social origins of the German sample are consistent with the claim.

7. The same is true of the proportions who attended selective secondary schools in the two countries, taking the data from the same two surveys (see Table 10.2).

TABLE 10.2

	Dipl. Ing. / graduate engineers per cent	Ing. Grad. / non-graduate engineers per cent
UK attended grammar or public school	83	40
BRD attended *Gymnasium*	93	53

It is only a small point but it is interesting to see that in Germany relatively more people with the advantage of a *Gymnasium* education viewed engineering as a worthwhile career.

8. There is an interesting suggestion too, when placing the data from the two surveys side by side, of a greater loyalty to engineering than to management in the German case. The questionnaire item invited respondents to say whether at the time they decided to become engineers they thought of an eventual career in management or only of working as an engineer (they were not forced into an either/or choice so the total is well under 100 per cent (see Table 10.3).

TABLE 10.3

At time of deciding to become an engineer	UK per cent	BRD per cent
thought of working as an engineer	40	50
thought of working (eventually) as a manager	25	14

9. The same sources also suggest something of a narrower subject loyalty where the Germans are concerned (see Table 10.4).

TABLE 10.4

	UK per cent	BRD per cent
studied mechanical engineering subjects only at college	44	74

10. Not only is there this greater German loyalty to mechanical engineering per se as the main subject of study; the Germans are also more strongly attracted to technical subjects generally (see Table 10.5).

TABLE 10.5

	UK per cent		BRD per cent	
	graduates	non-graduates	Dipl. Ings.	Ing. Grads.
studied non-technical subjects at college	41	72	24	12

The 'top' non-technical subject in Britain was Industrial Administration; in Germany it was *Betriebswirtschaft* (business economics). These course preferences and career perceptions are all in the German case in line with the dominance of *Technik* and specialization norms. This last contrast, regarding the proportions who studied non-technical subjects as part of their course, is striking, and the difference in Britain's favour would presumably be even greater if the British survey had been done at the later date of the German.

11. Looking back on their college education very similar proportions expressed satisfaction with this earlier training. A difference does emerge, however, with regard to a retrospective evaluation of the *practical* training in industry which they had received: a much higher proportion of the Germans expressed satisfaction with this.

12.   In both surveys the respondents were invited to construct an ideal engineering course: to specify the main inputs, that is, and give them a proportional weighting. This exercise produced some sharp differences (see Table 10.6).

TABLE 10.6

	UK	BRD
		per cent of ideal course
design	13	24
speciality engineering	6	11
basic engineering science	27	3

This again affords a small piece of evidence for German specialization. The greater importance attached to design here is consistent with our experience in German companies. Design tends to be more prominent in a German company; it is not unusual to hear Design described as a *prima donna* function, and it appears that a higher proportion of all German engineers actually work in Design.

13.   A higher proportion of the German graduates (10 per cent) took doctors' degrees; the corresponding figure for the British sample was 3 per cent, though this proportion might be higher today.

14.   More of the Germans had published books and articles (see Table 10.7).

TABLE 10.7

UK non-graduates per cent	BRD Ing. Grads. per cent	UK graduates per cent	BRD Dipl. Ing. per cent
4	25	10	49

15.   Similarly more of the Germans had given lectures or presented papers outside their place of work (see Table 10.8).

TABLE 10.8

UK non-graduates per cent	BRD Ing. Grads. per cent	UK graduates per cent	BRD Dipl. Ings. per cent
10	32	11	52

With the last two items it is possible that, had the survey in Britain been carried out later, the British showing would be better. On the other hand the difference as it stands is enormous, and even with the qualification about the lapse of time between the two studies, this difference is probably a reflection of the rigour and standing of German engineering courses, and the apparently whole-hearted engagement of the German engineer in *Technik*, which several other items in this summary section have also suggested.

16. More of the Germans had produced patented inventions (see Table 10.9).

TABLE 10.9

UK non-graduates per cent	BRD Ing. Grads. per cent	UK graduates per cent	BRD Dipl. Ings. per cent
12	23	8	26

Even allowing for any possible effect of the time gap between the two studies, this contrast will come as a surprise. It is frequently claimed that originality and inventive ability are British strengths, even if our record on successful commercial exploitation is less impressive. There may be a partial explanation in terms of different national attitudes to patenting. There is no doubt in our minds that Germans like patenting: it accords with their pride in *Technik*, specializing norms, and clear desire to make the most of these, commercially speaking. In Britain, on the other hand, the view is sometimes expressed that patenting an invention is dysfunctional: it advertises the invention and anyone can find a way round the patent laws by introducing an inessential modification. Even with this interpretative modification, however, the difference is quite large.

17. Changing direction a little it is interesting to note that more of the German engineers work in industry (see Table 10.10).

TABLE 10.10

	UK per cent	BRD per cent
First job in industry	75	80
Present job in industry	64	71

This difference is not large but it is consistent with the Anglo-German industry-*vis-à-vis*-public sector pay relativities explored earlier, as well as the higher standing claimed for German industry.

18. Relatively more of the German engineers worked in smaller organizations (see Table 10.11).

TABLE 10.11

	UK per cent	BRD per cent
working in organizations with fewer than 100 employees	7	21
working in organizations with more than 500 employees	74	54

Again the British showing on engineers employed in relatively small companies would probably be (a little) higher today, but the difference as it stands is quite a large one. It is entirely consistent with our experience of German industry where quite small firms turn out to have engineers qualified at the Ing. Grad. and Dipl. Ing. level employed immediately above the foreman level in the company structure.

19. A higher proportion of the German engineers had the use of a secretary (72 per cent against 43 per cent of the British sample). This small point doubtless reflects the German engineer's greater access to general-management posts in industry, but it should not be over-interpreted since it is probably also conditioned by greater German wealth — when it comes to resources and facilities the Germans have more of them anyway.

20. The German engineers changed jobs much less frequently than their British colleagues. In both cases all-age samples were surveyed: on average the German engineers had 1.9 jobs, and the British 2.7. At the time of the survey 36 per cent of the German engineers were still in their first job as opposed to 16 per cent of the British sample. Our interpretation is that this reflects the greater contentment and satisfaction of the German engineers, but differences of career style and expectations may also be involved.

21. A particular instance of such possible differences in style is the relation between job mobility and occupational success (as defined in Chapter 5). In Germany there appeared to be no correlation between success and mobility, though such a correlation did exist in Britain.

22. The German engineers exhibited a higher level of job satisfaction. To take a simple indicator 36 per cent of the German sample described themselves as 'satisfied in all respects'; whereas 16 per cent of the British sample described themselves as being without major dissatisfaction or frustration. A higher proportion of the Germans also said that they would choose engineering again if they were starting their careers once more, but the difference was only small.

23. The Germans appear to have longer hours of work (actual rather than prescribed), the average being 48 per week; the corresponding figure for the British sample was 42 per week including work done at home. The strictly comparable German figure may thus be higher since the question put to the German engineers did not allow for work at home.

The foregoing is not intended as a comprehensive summary.

The idea is rather to re-group some of the thematically relevant data, and to show a little bit by contrast what the *Technik* society is like.

The Imponderables of Learning

If it has been established that there is a sufficient gap between Britain and Germany in these respects for the question of learning to be raised in principle, this is not quite the same as showing that learning is actually feasible. In fact the question, can one country learn from another, is rather difficult to answer. This is partly because there are some genuine imponderables and partly because the seemingly simple question is really several questions interwoven.

The question of whether one country can learn from another is perhaps best answered by looking round to see if it has actually happened. Perhaps the best example of learning by international borrowing on a wide front is Japan, a country which stage-managed its transition into modern industrial statehood on the basis of extensive borrowing from the West. There are also many examples of international learning in particular spheres. In the period before the First World War Turkey and several Latin American states learned military organization and tactics from German military missions, and more recently African insurgents have learned guerrilla operations from the Cubans. The British learned management theory and practice from the Americans, and the Germans took lessons in classlessness from the Americans at the end of the Second World War. Mother countries have also variously inspired and moulded their colonies. We British have spread cricket, constitutionalism, and precedental law systems around the world. The French succeeded magnificently in transmitting their model of sophistication and intellectual panache to the élites of their African and Asian colonies. Indeed it is rather a pity, from the point of view of understanding the dynamics of international culture exchange, that the Versailles Treaty of 1919 deprived Germany of its few colonies. If this had not happened we might today be able to point to the Tanzanians as the punctuality and performance obsessed *Technik* lovers of the African continent.

If we can be confident that the world has some instances of

cultural borrowing, there are still some other inscrutables. One can never be sure what has caused what in the model country, and this is especially true with regard to end states. If one is concerned, for instance, with the German economic performance, and with emulating it, it is difficult to demonstrate in any conclusive way just how much German engineering has contributed to the performance. Or to take another example, it may be that our analysis of the determinants of the German engineer's status is all wrong; it is logically possible that the *real* cause is something we have not thought of. Similarly, even if one has identified a desirable model and understood correctly its internal dynamics there is still no way of knowing whether some foreign institution or arrangement will work at home. Perhaps there is some cultural barrier to its transfer which has not previously been identified. It is important to underline the word perhaps in the previous sentence: the existence of such barriers is not invariable, and the argument should not be treated as any kind of an absolute.

The effect of these imponderables is to change the way in which any decision to adopt some practice from a foreign country has to be taken. In the end this has to be done on the basis of common sense and probability, rather than logical rectitude and certainty. There is also a corresponding freedom in these discussions of learning from other countries. It is that one does not need to restrict oneself to the tangible and administratively transferable items: the relative merits for instance, of longer or shorter courses, the compulsory inclusion of the *Praktikum*, whether or not engineering lecturers should be obliged by law to have worked in industry first, and so on. These may be significant and often represent the way things regarded as important are 'put into operation' in another country, but they are not the whole story. There are also the attitudes and beliefs themselves, and other aspects of the foreign country that appear to be conducive to the performance for which the country is respected.

In short the present writers are not a two-man royal commission charged with producing instantly applicable recommendations. We are free to speculate about possible specific changes, but also to indulge in a little comparative cultural analysis if this has background relevance.

Starting at the End

There is an old saw in industry: what would you do if it was your money? A variation on this might be: what would you do if this were for real? In other words, if there is going to be a discussion about learning from another country, to what *end* is this to be done. What does one want to achieve?

It is a key question. We do not want to emulate Germany, or anywhere else, for trivial reasons, for fun, or indiscriminately. So does West Germany have anything we want?

West Germany has greater wealth. It is richer as a nation; it has in consequence more options, more room for manoeuvre, more to spend, and a greater actual freedom to pursue quality-of-life objectives. For most Germans there is also a much higher level of personal wealth, with the freedom, and sometimes freedom from worry, that this implies. Or, to return again to the impersonal level, West Germany is *relatively* free from the endless economic problems which have and do beset Britain. There is no point in denying that the Germans are quite shamelessly attached to this wealth. There is a saying in Germany: with money you can't buy happiness, but it is a great down payment.

These claims would not hold for Germany if that country had not developed and sustained an effectively functioning manufacturing sector. Before working out the relation between engineering and this achievement in the German context, a sideways look at Britain is in order.

British society has not been hospitable to industry, and has for the most part left industry to 'make out' as best as it can. In this situation industry has, with regard to any foreign learning, taken the line of least resistance in a cultural–linguistic sense. It has, that is to say, taken the USA rather than continental Europe, as the exemplar of business practice. The inspiration from this source, however, has concerned motivation and organization rather than the role of *Technik* in the creation of national wealth. Industry in Britain, given its very lukewarm socio-cultural reception, has also been forced on to the defensive. In seeking a *raison d'être* it has seized the most comprehensible, the most tangible, and conceivably the most misleading rationale — profit.

Business as usual in West Germany

In sociological writing in the last decade or more a recurrent idea is that of the *post* industrial society. A constant element in these formulations is the putative shift of the occupational centre of gravity from the secondary sector (manufacturing) to the tertiary sector (services), with the emphasis on thinking, knowledge-creating, and facilitating, rather than on actually producing. Tempting as it would be to stigmatize this as a piece of arrant British snobbery (the anti-industry, anti-work, and anti-production bias is all too clear) it has to be admitted that it all started in America.[4] One of several reasons for being grateful to the Finniston Committee is for its lucid repudiation of this view that national wealth can be obtained without actually making anything.[5] Manufacturing simply accounts for too large a proportion of the GNP to be regulated in this way. What is more, the Germans have shown an intuitive grasp of this. Theirs is the most unashamedly industrial of all societies. They do not expect to make a living from invisible exports or tourism, by business manoeuvres or brilliant marketing, by pricing policy or creative advertising: they proceed on the assumption that if a product is well designed, well made, and finished on time, then the rest will follow.

Germany and Technik

The German approach to business is simple, and so far it has worked. It may be worth exploring a little further the role of the engineer and the *Technik* culture in this enterprise.

Engineers are absolutely and relatively more numerous in Germany, and their dominance of industrial management is that much stronger. To repeat in short the testimony of an earlier chapter, engineers in Germany dominate completely the technical functions in industry, they 'overspill' into the non-technical functions, and they are very well represented, if only by sheer weight of numbers, in higher general management. All this means a powerful standing lobby for design, production, and quality. It is natural, that is to say, for them to espouse these concerns, to comprehend the initiatives of others in these spheres, and to respond. It means that German businesses are often in the words of their managers, *technisch gesteuert* (directed and managed in terms of the dictates of *Technik*), and so quite often are non-technical functions like Sales and Advertising.

This stronger technical orientation of German companies generally means that the technical functions tend to have higher standing than in Britain. This is noticeable in several ways. One which was mentioned earlier is that Design is more generally esteemed. People in Design are regarded as creative in a positive way, and there is less gratuitous ascription of impracticality to them than there is in Britain. The difference also shows in quality matters. Quality is far too important to be left (exclusively) to Quality Control. It is quite normal for German production managers to intervene in quality matters on their own initiative, to take preventive or corrective measures or conduct little investigations without any stimulus from Quality Control. In periods spent as an observer in German companies one of the present authors has witnessed several instances where production managers jointly planned quality audits and final inspection procedures with colleagues from Testing and Inspection, and one occasion on which a production manager staged a confrontation with the quality control manager on the grounds that the latter's checks and monitoring were not strict enough.

The manufacturing or production function is also a beneficiary of this *Technik* culture, though care is needed in defining any Anglo-German difference here. In Britain there has been a tendency to regard Production as a 'Cinderella' function, one which ambitious and well-qualified managers will avoid, even if the tide is now turning in Production's favour. There are echoes of this in Germany too, though the feeling is less widespread. But what is remarkable to an observer of the German company scene is the confidence and authority with which production managers conduct themselves in their dealings with other managers. In short, German production managers intervene in production-related areas at will, and treat other functions as resources to be summoned and dismissed at will.

The corporate background for this kind of deportment by production managers is quite simply a very product-oriented one. German managers 'talk products' and manufacturing more than their British colleagues do, and this applies to

German managers generally, not just those associated with design and product development. An easy way of putting this claim to the test is to ask senior German managers what, as representatives of their company, they are most proud of. The answers tend to emphasize products, product quality, design, manufacturing techniques, and after sales service, and to de-emphasize profit, turnover, market share, and acquisitions.

One effect of the engineers' dominance is that there is more homogeneity of qualification in the technical functions, and probably more mobility between them. In all the technical functions anyone who is anyone is an Ing. Grad. or Dipl. Ing., and there is a leaven of Dr. Ings. (Ph.D.s in engineering) as well. This in turn tends to lower the boundaries between these functions with regard to mobility between them. Thus in our experience there is a lot of interchange between Engineering, Production, Maintenance, and Quality Control. It is also noticeable that a lot of German production managers seem to have worked in Design at an earlier stage of their career. And a concomitant of this mobility and qualificational homogeneity is that there appears to be less (hostile) stereotyping. Again, as was suggested in an earlier chapter, the idea of *Technik* tends to bring together all those concerned with manufacturing at different levels: it serves, that is, the end of vertical integration as well.

We have tried here to show some of the positive effects of the concentration of qualified engineers in German manufacturing industry, and of the *Technik* ethos. None of this, however, is taking place in a socio-cultural vacuum. In earlier chapters occasional reference has been made to other aspects of German society, and it is worth recalling three of these. They constitute part of the ethos of German industry, and although they are in no way the exclusive property of engineers, they are often typified by engineers.

First Germans exhibit a marked penchant for specialization, and, as was noted earlier, this reinforces the position of the engineer, who gives it an important application. This valorization operates as well on a much wider front to underpin vocational education in general. In the German scheme of things there is a marked emphasis on specific knowledge and experience, and this plays a major role in the attempt to match people to jobs. If one talks to German personnel and training officers in industry their pronouncements tend to be more particularized than those of their British colleagues. Naturally they make some reference to general personality considerations but they are often able to state job requirements precisely in terms of the level of qualification, content of qualification, and particular types of experience. In the German mind specialism is counterposed to amateurism as much as to generalism. In the sphere of manufacturing it is conducive to a concentration on products, design, methods, quality, and reliability.

The second feature is the relative classlessness of German society. While there are formal differences of wealth and power, educational level, and job status in Germany, as elsewhere, there are far fewer behavioural and stylistic differences. There is less class-conciousness in everyday life, less tendency to classify people in social-class terms, and fewer ostensible differences on which to base such classifications.

An interesting speculation is how the Germans 'got this way'. It is possible, of course, that Germany was never as class-conscious and class-differentiated as Britain. It may also, in part, be a legacy of the national-socialist period, which had the effect of setting at nought the traditional allegiances and statuses. Very naturally the Third Reich has been more often subject to moral condemnation than sociological analysis: the fact remains that it may have been socially levelled by fear and at the same time inaugurated a demoniac *carrière ouverte aux talents*.

Germans themselves frequently offer a different explanation for classlessness-as-social unity. This is that the miseries of the whole German population at the end of the War were such as to obliterate (former) class and status differences, that further and subsequently, the whole population were involved in the work of reconstruction, and this had a profoundly unifying effect on the population. A sub-theme in this German penchant for relating classlessness to the Occupation period, is to allege that the American occupying troops set a practical example of classless egalitarianism. Whatever the determinants of German classlessness it is clearly a factor in their relative industrial peace, in the absence of unnecessary friction, in freedom from disruption, and thus in industrial output.

The third feature of German society, and it has come up several times already and especially in the earlier discussion of occupational success, concerns the national attachment to *Leistung* (performance, achievement). Again it is difficult to establish the reasons and stages of German evolution in the direction of the *Leistungsgesellschaft* (achievement society), but if the aetiology is unclear the popular attachment to the idea is not in doubt. And the sphere of German society in which devotion to *Leistung* is most conspicuous is again industry. This has a number of practical implications.

Leistung is a concomitant of classlessness, in the sense that it, rather than birth, or any ascriptive factor, is the determinant of personal standing. The attachment to *Leistung* is an antidote to privilege, though privilege does not appear to loom very large in the German placement system. It is quite compatible as well with a valorization of the engineer. Although German engineers have a lengthy training and an impressive record on what might be termed quasi-academic indicators —technical reading, publications, foreign language knowledge, and so on — the German view of the engineer is of a *Könner*, of someone who can *do* something, of someone at least in general *leistungsfähig* (capable of achievement).

The attachment to *Leistung* also makes Germans impatient of problems. In the German view problems are not there to be talked about, for ontological interest, or to enhance the individual's perceived interiority: problems are there to be solved (often by putting them under the spotlight of specialist knowledge), eliminated, 'rolled up'. And a particular manifestation of this achievement orientation in German industry is the conviction that things have to be achieved on time.

Circles

It will be clear from the foregoing discussion that the German system is self-fuelling. The economy or more particularly the manufacturing sector has certain strengths; so far these have led to success; this confirms the Germans in their system, priorities, and emphasis.

Put another way a 'virtuous circle' is in operation in Germany. The status of industry, status of engineering, concept of *Technik*, quality of manufacture, and economic success (especially the export of manufactured goods) are all mutually reinforcing. Or again, at a lower level, the quality of engineering education, access of engineers to management posts, status of industry, remuneration of engineers, and quality of new recruits to the occupation are all aligned in a circular interdependence.

Unfortunately it is not possible to claim the existence of such virtuous circularities for Britain, where, if anything, they are inverted. The question is how and where do we break in to the British circle of low status and under-achievement? The view we wish to urge here is that the optimal point for constructive intervention is the education system. This is argued partly because the realm of education does offer some real opportunity to effect change, and partly because it is one of the few areas susceptible to direct government intervention. To explore for the moment the negative side of this equation one cannot have status by statute, abolish 'applied science'; introduce *Technik*, *Leistung*, or classlessness by legislation; produce an 'engineering culture' by executive means; or force industry to raise the pay of engineers, the status of Design, or the number of production representatives on the board of directors. So our choice is to try to stimulate change via the education system.

Arguably the most radical of the proposals contained in the Finniston Report[6] is the suggestion that undergraduate students of engineering should benefit from differential grants; in fact a modest £250 above the statutory grant (determined by parental income), spread over three or four years, at 1980 prices, is proposed. While the motivational effect of a pound a week is doubtful, the basic idea of differential grants would appear to have considerable potential.

Suppose that differential grants were introduced in favour of engineers but with a very substantial differential — one which would make engineering undergraduates the aristocrats of the student world. It is likely that this would raise the quality of the intake in terms of 'A' level grades (tangible and measurable) and in other ways; it would give admissions tutors the chance to pick and choose, to discriminate and raise standards; it would soon redress the QSE ratio in favour of engineering; and, it would condition the expectations of both engineers and potential employers about remuneration.

There is a case for making the four-year degree course standard. This would go some way towards bringing our courses into line with those of the EEC and continental Europe generally. It would facilitate the inclusion of episodes of industrial training and non-technical subjects. The primary reason for increasing the length of the undergraduate course, however, must be to deepen the students' knowledge of engineering and enlarge his ability to practise engineering. If engineering is worth doing at university it is worth doing well.

Within the ambit of a longer undergraduate course there is a dual justification for the inclusion of some non-engineering elements, though as was made clear earlier in the chapter this is not something we need to learn from Germany since we already have the lead in this direction. The inclusion of such elements may render the graduate more attractive to future employers, especially in so far as such non-technical course components are designed to produce engineer-managers. The second consideration is that rightly or wrongly, and in Chapter 7 we argued on the basis of German evidence that it is wrongly, there is a tendency to view engineers as narrow and lacking in non-technical interests. The regular inclusion of some non-engineering elements in the undergraduate course will militate against this charge.

A compulsory *Praktikum* (period in industry), either before or during the undergraduate course (or both), is to be recommended for its own sake. As with the inclusion of non-technical course components it has also a tactical value. The feeling in Britain that the university-educated engineer will be too theoretical, too little able to apply his knowledge, and quite probably too conceited to boot, runs very deep: it is an attitude which must be combated, and a system of undergraduate engineering education which guarantees some direct experience of industry is a contribution to this end.

It may be noted that the suggestions offered here differ from those of the Finniston Report in that practical training and industrial experience are recommended as prerequisites or components of the undergraduate course, rather than as appendages to it. There is nothing absolute about this preference but it is only fair to make our reasoning clear. We suspect that large-scale post-graduate training in industry is difficult to organize, difficult to monitor, and very difficult to standardize. This substantial post-graduate requirement may also direct engineers even more to the largest firms and groups, conceivably to the detriment of small and medium-sized companies. This would be a pity. We know already that qualified engineers are to be found lower down the company-size hierarchy in Germany, and it is arguable that the firms which would benefit from the services of a highly-trained engineer are those who have never had one before, rather than the corporate giants. There is another consideration. The case for industry conducting a substantial post-graduate training operation is clear: it ought to guarantee that industry produces what it believes it wants. There is a corresponding argument against: it is that industry may not know what it wants or may be mistaken in its perceptions. In any case, to let it mould too closely the eventual intake into its full-time ranks will lessen the likelihood that it would receive any new impetus from its recruits.

Most of the reference so far has been to training at university level. Without formulating a precise proposal, there is a model which might be offered. We do not mean at this stage to launch into an encomium on the present German Ing. Grad. system. Rather we would like to draw attention to some aspects of the old Ing. Grad. system which the Germans themselves scrapped some ten years ago.

To repeat a little the account in Chapter 2, the earlier system worked in the following way. The eventual Ing. Grad. would leave school usually at the statutory leaving age, or sometimes a little later with the German equivalent of 'O'level, and do a normal apprenticeship in industry lasting some three and a-half years. Thereafter a short qualifying course (weeks not months) admitted one to the *Ingenieurschule* (the forerunner to the *Fachhochschule*) where a practically oriented three year full-time course, plus another six month *Praktikum*, led to the Ing. Grad. award. Such a system seems to offer certain advantages.

To begin with a small consideration, it in no way disadvantaged aspirants from a working-class background. Indeed the old system served to accommodate and even encourage them. Two German writers on the changes in the Ing. Grad. system have noted that whereas the Ing. Grad. course used to be a strong choice for able working-class lads it has now become a weak choice for middle-class lads who lack the drive to get to university.[7] Secondly, the old system had the further advantage that it did not militate against those who, for whatever reason, wanted to leave school at the statutory leaving age, and do something active and practical rather than continuing in some form of further education. But the old system did offer them the opportunity to re-enter full-time education later. Furthermore this system of the apprenticeship as a prerequisite for the *Ingenieurschule* (a two year supervised practical in industry used to be an alternative to the apprenticeship though most *Ingenieurschule* entrants came via the apprenticeship route) guaranteed that the later Ing. Grad. had several years practical experience in industry, a social familiarity with the factory as a place of work, and had demonstrated some craft skill.

To these considerations should be added the fact that the Ing. Grad. course was, and is, a full-time course. We do not mean with this remark to set up some snobbery about full-time courses being 'better' than part-time ones. The consideration is quite practical, and it is that part-time courses may be a very inefficient way of transmitting knowledge and developing skill. Indeed the much vaunted ONC-HNC course in Britain may have derived some of its lustre from being a test of character and endurance, rather than from its purely engineering excellence. There is another disadvantage to the part-time course. This is that from the viewpoint of the employer it has no discernible effect. Consider the ONC-HNC again. An earnest and enthusiastic lad starts a part-time course at the local technical college when he is 15 or 16. If he is good, works hard, and has endurance, some 5 or 6 years later he actually gets the HNC. But does his firm notice? Is he instantly upgraded? Does he get new employment as a qualified engineer? Whereas under the Ing. Grad. system, old or new, the Ing. Grad. comes on to the labour market with a newly acquired qualification and must be placed in terms of it. There can be no question of his becoming an over qualified lab. technician or second assistant to a rate fixer. Finally, with regard to the Ing. Grad. the lecturers on this course are all university graduates *and* they are required by law to have a minimum of five years' experience in industry. This double claim could not be made for the entire teaching body on any of the courses in Britain which roughly equate to the Ing. Grad. course in Germany. We would like to suggest that something may be learned from this (now discarded!) German model. Engineers who are trained at university are not the only ones, nor the only ones that matter. It is important to have the best possible training for those at the next level.

This point is already implicit in the suggestions that have been made but should be made explicit. There is a strong case for bridging the gap between the worlds of industry and engineering education. There is no shortage of co-ordinating committees and consultative bodies: what is needed, quite simply, is a bringing together of the two spheres by an overlap of personnel and experience. Germany is not the only model, but this bridging has been done quite satisfactorily in that country.

In these suggestions we have concentrated on the education system for the reason given at the beginning — it is the area in which the government can directly intervene. There is another consideration. If groups are chosen for ability and ambition, nurtured in specialized knowledge and aptitude, and given indeed some of the hallmarks of the graduates of the *École Polytechnique,* perhaps they will come to embody a German style commitment to *Leistung*. We need it.

Notes

CHAPTER 1

1. I. A. Glover, 'The Sociological and Industrial Relations Literature on Engineers and Engineering', Report to the Department of Industry (London, 1973).
2. For an early statement of this kind see M. J. Fores, 'Engineering and the British Economic Problem', *Quest*, No. 22 (Autumn 1972).
3. *Education Engineers and Manufacturing Industry*, a Report to the British Association Co-ordinating Group (Aug. 1977).
4. The *Guardian* newspaper and the *Financial Times* are examples, and among the engineering and management journals there have been such articles and series in *Engineering Today, Management Today, CBI Review, Business Graduate, Energy World, Chartered Mechanical Engineer, Chemical Engineer, Engineering,* and *Design Magazine*.
5. This emerges clearly in the British study discussed in Joel Gerstl and Stanley Hutton, *Engineers: the Anatomy of a Profession* (Tavistock Publications, London, 1966).
6. Eugon Kogon, *Die Stunde der Ingenieure* (VDI Verlag, Düsseldorf, 1976).

CHAPTER 2

1. For more detailed information see *Bildungswesen im Vergleich* (Bundesminister für Bildung und Wissenschaft, 1974), an excellent comparative study of the educational system in Germany commisioned by the German Minister for Education and published in six booklets; also *Engineering Education and Europe* (British Council, 1974), a comparative description of the tertiary educational systems for engineers in EEC countries; and Prof. A. W. J. Chisholm, 'First Report on the Education and Training of Engineers on the Continent of Europe' (University of Salford, 1975), a comparative study of Germany and France.
2. The Secondary Modern schools were part of the English system up to 1965, since when most of them have been superseded by Comprehensive schools. But as there has been no comparable development in Germany, we will use this outmoded, but more specific nomenclature.
3. In Germany the university engineering schools have been separate from the universities where one could read anything but engineering. Originally therefore an engineering school of university level was called *Technische Hochschule* which is translated as technical university or technological university. More recently, following educational fashion, many of these have adopted the title *Technische Universität*, literally technological university (TU for short), which makes their status quite clear. We shall therefore refer to all these engineering schools in English

as 'University' or 'Technical University' and try by emphasis and context to make it clear when 'university' is used in the general British sense covering all disciplines.

4. *Ingenieurschule* is the old form of the *Fachhochschule*: it is at (both) these institutions that the non-graduate engineer, the Ing. Grad., is trained.
5. See note 3 above.
6. B. Lutz and G. Kammerer, *Das Ende des graduierten Ingenieurs* (Europäische Verlagsanstalt, 1975).
7. Bund-Länder Kommission für Bildungsplanung und Bundesanstalt für Arbeit *Studien und Berufswahl* (Karl Heinrich Bock, Bad Honnef, 1974).

CHAPTER 3

1. J. E. Gerstl and S. P. Hutton, *Engineers: the Anatomy of a Profession* (Tavistock Publications, London, 1966).
2. J. Nimmergut, *Deutschland in Zahlen* (Wilhelm Heyne Verlag, München 1974).
3. H. Bartenwerfer and H. Giesen 'Pilotstudie über die Beobachtung und Analyse von Bildungslebensläufen' (Deutsches Institut für Pädagogische Forschung, Frankfurt, 1973).
4. Liam Hudson, *Contrary Imaginations* (Methuen, London, 1966).
5. J. E. Gerstl and R. Perrucci, *Profession without Community: Engineers in American Society* (Random House, New York, 1969).
6. Herman Bayer and Peter Lawrence, 'Engineering Education, and the status of Industry', *European Journal of Engineering Education*, No. 2 (1977).
7. Ibid.
8. L. Grinter *et al.*, 'Final Report on evaluation of engineering education', *Journal of Engineering Education*, 46 (1955) pp. 26–60.
9. S. P. Hutton, 'Opinions on the Social Science Content of the ideal engineering course', *Proceedings of the Zweiter Internationaler Kongress für Ingenieurausbildung* (Technische Hochschule, Darmstadt, Oct. 1978).

CHAPTER 4

1. 'I Mech E Salary Survey 1976', Supplement to *Mechanical Engineering News* (Mar. 1976).
2. J. E. Gerstl and S. P. Hutton, *Engineers: the Anatomy of a Profession* (Tavistock Publications, London, 1966).
3. S. P. Hutton, 'The Engineering Ph.D. is alive and well in Britain', *New Scientist* (23 Mar. 1978), pp. 792–3.

CHAPTER 5

1. In this and the following table English equivalents of German qualifications are used for convenience, but the German education system is described in some detail in Chapter 2.
2. These are proportions calculated with reference to the whole sample, not just to its Dipl. Ing. (graduate) members.
3. This last point is made clear in J. E. Gerstl and S. P. Hutton *Engineers: the Anatomy of a Profession* (Tavistock Publications, London, 1966).

CHAPTER 6

1. Sir Montague Finniston (Chairman), *Engineering our Future,* Report of the Committee of Inquiry into the Engineering Profession (HMSO, London, 1980).
2. G. Hortleder, *Das Gesellschaftsbild des Ingenieurs* (Suhrkamp Verlag, Frankfurt, 1970).

CHAPTER 7

1. One of the present authors has discussed elsewhere the grasp and originality of Simmel's characterization of these social roles: P. A. Lawrence (ed.), *Georg Simmel: Sociologist and European* (Thomas Nelson, London, 1976).
2. Eugon Kogon, *Die Stunde der Ingenieure* (VDI Verlag, Düsseldorf, 1976).
3. G. Hortleder, *Das Gesellschaftsbild des Ingenieurs* (Suhrkamp Verlag, Frankfurt, 1970).

CHAPTER 8

1. In Germany university courses, including engineering courses, are of longer duration than they are in Britain, but one has some choice in when one presents oneself for the final examination. Successfully completing the course in a shorter-than-average time is something that may be quoted with pride, like getting a first. Nine semesters equals four and a-half years and is below the national average for an engineering degree.
2. This argument, to the effect that the modern German education system reflects and contributes to the status of German industry, has been developed in more detail in an earlier book by one of the present authors: Peter Lawrence, *Managers and Management in West Germany* (Croom Helm, London, 1980).

3. See particularly Arndt Sorge and Malcolm Warner, 'Manufacturing Organisation and Work Roles in Great Britain and West Germany', Discussion paper of the Internationales Institut für Management (West Berlin, 1978).
4. The date is not significant except that our German data became available at the same time. The same pattern emerged from the same British survey in the following year, and is the same as that produced by the comparable survey of electrical engineers in Britain. We have no reason to suppose that it has changed.
5. Some of this material has also been discussed in Peter Lawrence, *Managers and Management in West Germany* (Croom Helm, London, 1980).
6. The *Vorstand* is the senior executive committee at the head of the large AG type of public company in West Germany.

CHAPTER 9

1. *Engineering Our Future,* Report of the Committee of Inquiry into the Engineering Profession, chaired by Sir Montague Finniston (HMSO, London, 1980).
2. This is argued for instance in Eugon Kogon, *Die Stunde der Ingenieure* (VDI Verlag, Düsseldorf, 1976), where the author considers the results of previous surveys.
3. *Education, Engineers and Manufacturing Industry,* a Report to the Co-ordinating Group of the British Association for the Advancement of Science (London, Aug. 1977).
4. The most detailed discussion of these German company data, with some British comparison, is in Hermann Bayer and Peter Lawrence, 'Effizienz und Traditionalismus: die Attraktivität akademischer Ausbildungsgänge in Grossbritannien und der Bundesrepublik Deutschland', *Soziale Welt,* Heft 1/2 (Münster, 1977).
5. M. J. Fores, 'Engineering and the British Economic Problem', *Quest,* No. 22 (1972).
6. Much of the survey evidence is discussed in Peter Lawrence, *Managers and Management in West Germany* (Croom Helm, London, 1980).
7. Washbos = Washington to Boston conurbation. Chipitts = Chicago to Pittsburg conurbation.

CHAPTER 10

1. J. E. Gerstl and S. P. Hutton, *Engineers: the Anatomy of a Profession* (Tavistock Press, London 1966).
2. Reported in a supplement to *Mechanical Engineering News* (Mar. 1976).
3. M. J. Fores 'Engineering and the British Economic Problem', *Quest,* No. 22 (1972).

4. Daniel Bell, *The Coming of Post Industrial Society* (Heinemann, London, 1974).
5. *Engineering Our Future,* Report of the Committee of Inquiry into the Engineering Profession, chaired by Sir Montague Finniston, (HMSO, London, 1980).
6. Ibid.
7. B. Lutz and G. Kammerer, *Das Ende des graduierten Ingenieurs* (Europäische Verlagsanstalt, Frankfurt, 1975).

Bibliography

BARNETT, CORELLI, 'Technology, Education and Industrial and Economic Strength', Cantor Lecture delivered to Royal Society of Arts, London, Nov. 1978.

BARTENWERFER, H., and GIESEN, H., 'Pilotstudie über die Beobachtung und Analyse von Bildungslebensläufen: Bericht über die Arbeiten des zweiten Projektjahres' (Deutches Institut für Pädagogische Forschung, Frankfurt, 1973).

BAYER, HERMANN, and LAWRENCE, PETER, 'Engineering, Education, and the Status of Industry', *European Journal of Engineering Education*, No. 2 (1977).

BAYER, HERMANN, and LAWRENCE, PETER, 'An emphasis on the practical in the land of idealism', *The Times Higher Education Supplement* (14 Jan. 1977).

BAYER, HERMANN, and LAWRENCE, PETER, 'Wirtschaftliche Effizienz und Traditionalismus: Die Attraktivität akademischer Ausbildungdgänge in Gross Britannien und der Bundesrepublik Deutschland', *Soziale Welt* No. 1/2 (Münster, 1977).

BELL, DANIEL, *The Coming of Post Industrial Society* (Heinemann, London, 1974).

BRINKMANN, G., *Die Ausbildung von Führungskräften für die Wirtschaft*, (Universitätsverlag, Micheal Wienand, Köln, 1967).

British Association Co-ordinating Group, *Education, Engineers and Manufacturing Industry* (London, Aug. 1977).

British Council, *Engineering Education and Europe* (London, 1974).

Bund-Länder Kommission für Bildungsplanung und Bundesanstalt für Arbeit, *Studien und Berufswahl* (Karl Heinrich Bock, Bad Honnef, 1974).

Bundesminister für Bildung und Wissenschaft, *Bildungswesen im Vergleich* (WEMA Institut, Cologne, 1974).

CHISHOLM, A. W. J., 'First Report on the Education and Training of Engineers on the Continent of Europe, with special reference to courses in total technology' (University of Salford, June 1975).

COHEN, L., and DERRICK, T., 'Occupational Values and Stereotypes in a Group of Engineers', *British Journal of Industrial Relations*, Vol. 8 (1970).

COTGROVE, STEPHEN, and BOX, STEVEN, *Science, Industry and Society* (Allen and Unwin, London, 1970).

Department of Industry (discussion paper), 'Industry, Education and Management' (London, July 1977).

Deutscher Verband Technisch-Wissenschaftlicher Vereine (DVT) (ed.), *Die Anforderungen des Berufs und die Ansprüche der Gesellschaft an den Ingenieur* (VDI Verlag, Düsseldorf, 1973).

DRUCKER, PETER F., *The Age of Discontinuity* (William Heinemann, London, 1969).

DUVE, FREIMUT (ed.), *Technologie und Politik* (Rowohlt Taschenbuch Verlag, Reinbek bei Hamburg, 1975).

FINNISTON, Sir MONTAGUE (Chairman), *Engineering Our Future*, Report of the Committee of Inquiry into the Engineering Profession (HMSO, London, 1980).

FORES, M. J., 'Engineering and the British Economic Problem', *Quest*, No. 22 (Autumn 1972).

FORES, MICHAEL, and GLOVER, IAN (eds.), *Manufacturing and Management* (HMSO, London, 1978).

FORES, MICHAEL, GLOVER, IAN, and LAWRENCE, PETER, 'Engineers in Germany', *Chartered Mechanical Engineer* (Oct. 1977).

FORES, MICHAEL, and LAWRENCE, PETER, 'Industrial Phobia', *New Society* (Oct. 1978).

FORES, MICHAEL, LAWRENCE, PETER, and SORGE, ARNDT, 'Germany's Front line Force', *Management Today* (Mar. 1978).

FORES, MICHAEL, LAWRENCE, PETER, and SORGE, ARNDT, 'Why Germany Produces Better', *Management Today* (Nov. 1978).

FORES, M., LAWRENCE, P., and SORGE, A., 'Fertigung ist ein schmutziges Geschäft' *Manager Magazin* (Apr. 1979).

FRAGNIERE, GABRIEL, and SELLIN, BURKHART, *Der Ingenieur in der Europäischen Gemeinschaft* (Hermann Schroedel Verlag, Hannover, 1974).

GERSTL, J. E., and HUTTON, S. P., *Engineers: the Anatomy of a Profession* (Tavistock Publications, London, 1966).

GERSTL, J. E., and PERRUCCI, R., *Profession without Community: Engineers in American Society* (Random House, New York, 1969).

GLOVER, I. A., *Professional Engineers and Engineering: A Review of the Sociological and Industrial Relations Literature*, Report to the Department of Industry (London, 1973).

GLOVER, I. A., and LAWRENCE, P. A., 'Engineering the Miracle', *New Society* (London, Sept. 1976).

GRINTER, L. *et al.*, 'Final report on evaluation of engineering education', *Journal of Engineering Education*, 46 (Sept. 1955), pp. 26–60.

HARTMANN, HEINZ, *Authority and Organisation in German Management* (Princeton University Press, Princeton, 1959).

HARTMANN, HEINZ, and WIENOLD, HANS, *Universität und Unternehmer* (Bertelsmann, Gutersloh, 1967).

HASLEGRAVE, H. L., (Chairman), *Report of the Committee on Technician Courses and Examinations* (HMSO, London, 1969).

HILMER, HOLGER, and LAWRENCE, PETER, 'Ingenieure im Vergleich: Mitglieder des VDI sind Elite beim beruflichen Interesse und beruflichen Bewusstsein', *VDI Nachrichten,* No. 43 (28 Oct. 1977).

HILMER, HOLGER, and LAWRENCE, PETER, 'Der britische Kollege', *VDI Nachrichten,* No. 44 (4 Nov. 1977).

HORTLEDER, GERD, *Das Gesellschaftsbild des Ingenieurs: zum politischen Verhalten der Technischen Intelligenz in Deutschland* (Suhrkamp Verlag, Frankfurt, 1970).

HORTLEDER, GERD, *Ingenieure in der Industriegesellschaft* (Suhrkamp Verlag, Frankfurt, 1973).

HUDSON, LIAM, *Contrary Imaginations* (Methuen, London, 1966).

HUTTON, S. P., 'Opinions on the Social Science Content of the Ideal Engineering Course', *Proceedings of the zweiter Internationaler Kongress der Ingenieurausbildung* (Technische Hochschule Darmstadt, Oct. 1978).

HUTTON, S. P., 'The Engineering Ph.D. is alive and well in Britain', *New Scientist* (23 Mar. 1978) pp. 792–3.

HUTTON, S. P. and LAWRENCE, P. A., 'The Mechanical Engineering Profession in Western Germany', *Conference Proceedings of the Institution of Mechanical Engineers* (London, 1976).

HUTTON, S. P. and LAWRENCE, P. A., 'The Engineer at School', *Energy World* (London, Feb. 1977).

HUTTON, S. P. and LAWRENCE, P. A., 'Getting Engineering back on the Right Lines', *The Times Education Supplement* (9 Sept. 1977).

HUTTON, S. P. and LAWRENCE, P. A., *Production Managers in Britain and Germany,* Interim Report to the Department of Industry (London, 1978).

HUTTON, S. P. and LAWRENCE, P. A., *The Work of Production Managers: Case Studies at Manufacturing Companies in West Germany,* Report to the Department of Industry and Science Research Council (London and Swindon, 1979).

HUTTON, S. P. and LAWRENCE, P. A., and SMITH, J. H., *Recruitment, Deployment, and Status of the Mechanical Engineer in the Federal German Republic,* Report to the Department of Industry (London, 1977).

'I Mech. E. Salary Survey 1976' *Supplement to Mechanical Engineering News* (Mar. 1976).

KAMMERER, GUIDO, LUTZ, BURKART, and NUBER, CHRISTOPH, *Ingenieure im Producktionsprozess* (Athenäum Verlag, Frankfurt 1973).

KOGON, E., *Die Stunde der Ingenieure* (VDI Verlag, Düsseldorf, 1976).

KRUK, MAX, *Die oberen 30,000* (Betriebswirtschaftlicher Verlag Gabler, Frankfurt, 1972).

LAWRENCE, PETER, 'The Comprehensive German', *Engineering* (London, Jan. 1977).

LAWRENCE, P. A., 'German lessons for non-graduate engineers', *The International Journal of Mechanical Engineering Education*, Vol. 5, No. 2 (Apr. 1977).

LAWRENCE, P. A., 'Engineers: the Image and Reality', *Energy World* (June 1977).

LAWRENCE, PETER, 'The German Engineer and the Status of Industry', *CBI Review* (Summer 1977).

LAWRENCE, PETER, 'An Engineer's Utopia', *Financial Times* (28 Oct. 1977).

LAWRENCE, PETER, 'Englische Ingenieure sind Stiefkinder', *Der Leitender Angestellter*, Ex. 27 (Nov. 1977).

LAWRENCE, PETER, 'Engineering and the Status of Production: West Germany', *The Business Graduate* (Nov. 1977).

LAWRENCE, PETER, 'Its the Product that Counts', *CBI Review* (Winter 1977/8).

LAWRENCE, PETER, 'Executive Head Hunting', *New Society* (May 1978).

LAWRENCE, P. A., 'The Engineer and Society', *Energy World* (May 1978).

LAWRENCE, PETER, 'Production Management: Problems and Remedies', *The Business Graduate* (Summer 1978).

LAWRENCE, PETER, 'Uber alles: engineers in West Germany', *Engineering Today* Vol. 2, No. 35 (2 Oct. 1978).

LAWRENCE, PETER, '*Technik*, a clue to German Product Success', *Design* (Dec. 1978).

LAWRENCE, PETER, 'Engineering and the National Context', *Electronics and Power* (May 1979).

LAWRENCE, PETER, 'Engineering Education and Meritocracy', *Higher Education Review* (Summer 1979).

LAWRENCE, PETER, 'Germany gives a guiding light', *New Scientist* (4 Oct. 1979).

LAWRENCE, PETER, 'The German approach to good management', *Engineering Today*, Vol. 3, No. 44 (27 Nov. 1979).

LAWRENCE, P. A., *Managers and Management in West Germany* (Croom Helm, London, 1980).

LUTZ, B., and KAMMERER, G., *Das Ende des graduierten Ingenieurs* (Europäische Verlagsanstalt, Frankfurt-Köln, 1975).

MAY, B., *The Engineer in Germany: A Review of the Literature* (Department of Industry, London, 1975).

MAY, B., *Social, Educational and Professional Background of German Management* (Department of Industry, London, 1975).

MCQUILLAN, M. K., *Graduate Engineers in Production* (Cranfield Institute of Technology, Cranfield, Aug. 1978).

NIMMERGUT, JORG, *Deutschland in Zahlen* (Wilhelm Heyne Verlag, München, 1974).

OPPELT, CLAUS, *Ingenieure im Beruf* (Max-Planck-Institut für Bildungsforschung, Berlin, 1976).

PETERS, R. W., and POLKE, M., *Beruf und Ausbildung der Ingenieure* (Verein Deutscher Ingenieure, Düsseldorf, 1973).

PRANDY, KENNETH, *Professional Employees: A Study of Scientists and Engineers* (Faber and Faber, London, 1965).

PROSS, HELGA, and BOETTICHER, KARL, *Manager des Kapitalismus* (Suhrkamp, Frankfurt, 1971).

ROTHWELL, ROY, 'Where Britain Lags Behind', *Management Today* (Nov. 1978).

SNOW, C. P., 'The Two Cultures and the Scientific Revolution', The Rede Lecture (Cambridge, 1959).

SORGE, ARNDT, and WARNER, MALCOLM, 'Manufacturing Organisation and Work Roles in Great Britain and West Germany', Discussion paper of the Internationales Institut für Management (West Berlin, 1978).

WEINSHALL, THEODORE D., *Culture and Management* (Penguin Books, Harmondsworth, 1977).

WILLIAMS, ROGER, *European Technology: The Politics of Collaboration* (Croom Helm, London, 1973).

ZAPF, W., *Die deutschen Manager-Sozialprofil und Karrierweg*, in W. Zapf (ed.), *Beitrage zur Analyse der deutschen Oberschicht*, 2nd. edn. (Piper, München, 1965).

Index